APPLICATIONS OF FUZZY LOGIC IN BIOINFORMATICS

Series on Advances in Bioinformatics and Computational Biology – Volume 9

APPLICATIONS OF FUZZY LOGIC IN BIOINFORMATICS

Dong Xu

James M Keller

Mihail Popescu

Rajkumar Bondugula

University of Missouri-Columbia, USA

ICP

Imperial College Press

Published by

Imperial College Press
57 Shelton Street
Covent Garden
London WC2H 9HE

Distributed by

World Scientific Publishing Co. Pte. Ltd.
5 Toh Tuck Link, Singapore 596224
USA office: 27 Warren Street, Suite 401-402, Hackensack, NJ 07601
UK office: 57 Shelton Street, Covent Garden, London WC2H 9HE

British Library Cataloguing-in-Publication Data
A catalogue record for this book is available from the British Library.

APPLICATIONS OF FUZZY LOGIC IN BIOINFORMATICS
Series on Advances in Bioinformatics and Computational Biology — Vol. 9

ISBN-13 978-1-84816-258-7
ISBN-10 1-84816-258-8

Printed in Singapore.

To our families

(left to right) Mihai, Dong, Jim, and Raj

Foreword

Bioinformatics is one of the youngest and most exciting fields in modern science. During the past decade, bioinformatics has become a challenging arena of applications of a wide variety of concepts and sophisticated techniques drawn from mathematics, computer science and probability theory. Within the fuzzy logic community, the meteoric ascent of bioinformatics has led to a contentious question: Can fuzzy logic make a substantive contribution to advancement of bioinformatics? The pioneering work "Applications of Fuzzy Logic to Bioinformatics," co-authored by Professors Dong Xu, James Keller, Mihail Popescu and Dr. Rajkumar Bondugula, may be viewed as a persuasive argument in support of an affirmative answer to the question. The core argument is that fuzzy logic is needed to solve problems in bioinformatics which are beyond the reach of existing techniques. It should be noted that "Applications of Fuzzy Logic to Bioinformatics," is the first book on this subject.

Today, fuzzy logic plays a relatively minor role in the armamentarium of bioinformatics. A metric is the number of publications with "fuzzy" in title or abstract—publications which are listed in the PubMed database. The current rate is 300-400 papers per year. Will the same be true in a few years from now? My belief is that in coming years there will be a rapid growth in the visibility and importance of fuzzy-logic-based techniques in the literature of bioinformatics and, more generally, in the literature of biological and medical sciences. However, my belief is based not on a detailed familiarity with bioinformatics—a familiarity which I do not have—but on my understanding of what fuzzy logic has to offer.

There are many misconceptions about fuzzy logic. Fuzzy logic is not fuzzy. Basically, fuzzy logic is a precise logic of imprecision. The principal objective of fuzzy logic is formalization/mechanization of imprecision, uncertainty, incompleteness of information and partiality of truth. Bioinformatics data fit this description, in addition to having a huge mass and high dimensionality.

Science deals not with reality but with models of reality. Concomitantly, scientific progress is driven by a quest for better models of reality. In bioinformatics, modeling is focused on genes, genomes and related biological entities. Brilliant successes have been achieved through the use of models based on bivalent logic and probability theory. However, there are many problems, such as those discussed in "Applications of Fuzzy Logic to Bioinformatics," in which better results can be achieved with better models based on the use of fuzzy logic. What is widely unrecognized is that modeling techniques based on bivalent logic and probability theory are intrinsically less powerful than techniques which are based on fuzzy logic and fuzzy-logic-based probability theory. An important contribution of "Applications of Fuzzy Logic to Bioinformatics" is making the bioinformatics community aware of the powerful modeling capability of fuzzy logic.

The superior capability of fuzzy logic as a modeling language is one of the principal rationales for its use in bioinformatics and, more generally, in scientific theories. An elaboration of this assertion is in order.

In a general setting, let $M(S)$ be a model of S. There are two basic metrics which might be associated with $M(S)$. First, the goodness of $M(S)$ as a model of S, call it cointension; and second, the computational complexity of $M(S)$. In general, cointension and computational complexity are covariant in the sense that an increase in cointension of $M(S)$ results in an increase in the computational complexity of $M(S)$. Bivalent logic and probability theory are, respectively, special cases of fuzzy logic and fuzzy-logic-based probability theory. What this implies is that, viewed as a modeling language, bivalent logic and probability theory have an intrinsically lower power of cointension than fuzzy logic and fuzzy-logic-based probability theory. However, the reverse is true so far as computational complexity is concerned. What gives fuzzy logic an advantage is that an increase in the computational complexity is far less important than an increase in cointension. This, in principle, is one of the main rationales for the use of fuzzy logic in bioinformatics. It should be noted that the relation between bivalent logic and fuzzy logic is similar in spirit to the relation between linear system theory and nonlinear system theory.

"Applications of Fuzzy Logic to Bioinformatics," serves three major purposes. First, it introduces fuzzy logic to the bioinformatics community. Second, it introduces bioinformatics to the fuzzy logic community; and third, it demonstrates that fuzzy logic has much to contribute to the advancement of bioinformatics. Professors James Keller, Dong Xu, Mihail Popescu and Dr. Rajkumar Bondugula, and the Imperial College Press deserve our thanks and congratulations for producing a work whose importance is hard to exaggerate. They deserve a loud applause.

Lotfi A. Zadeh
Berkeley, CA
September 24, 2007

Preface

Science is entering a new era thanks to the Human Genome Project, one of the largest programs in molecular biology. This project was devoted to the sequencing of human DNA fragments, i.e., to the determination of the order of nucleic acids therein. These sequences represent the blueprint of life. Since the 1980s, the advent of the Human Genome Project and other DNA sequencing projects has led to exponential growth in molecular data. Genomic sequencing has opened a new avenue to study biological systems on large scales, paving the way for investigating other high-throughput data. Today, due to the availability of high-throughput measurement technologies, it is possible to use a broad range of experimental data to expand the genome-scale studies from biological sequences and protein structures to higher-level functions and phenotypes. For example, microarray technology is a powerful tool to systematically measure gene expression across whole cells and tissues under varying experimental conditions or over a time course. As massive data are being generated, there is a strong demand for bioinformatics in data management, visualization, integration, analysis, modeling, and prediction. Bioinformatics has been developed extremely fast and has brought enormous impact to the research of biology and medicine in recent years. Thousands of bioinformatics databases and tools have been developed. More and more experimental biologists have realized the importance of bioinformatics, as the need for managing and analyzing the massive amount of data is evident. Many biologists now use bioinformatics tools themselves, especially through a Web interface.

As massive biological data have become a fundamentally important resource in biomedical sciences, researchers have developed various bioinformatics algorithms and software tools to identify meaningful information (or statistically significant patterns) from data and correlate such information for discovery of new knowledge or prediction of biological properties. However, such tasks are often highly challenging. The information-rich data are heterogeneous and ambiguous in nature. They are often noisy and incomplete, as well as containing misleading outliers. Furthermore, biological systems, due to adaptability, evolution, redundancy, robustness, and emergence, are extremely complex. The challenge has drawn a wide range of studies from computer sciences, and various computer science technologies have been applied. The most notable applications include dynamic programming, neural networks, hidden Markov models, support vector machines, etc. Fuzzy set theory and fuzzy logic have also been used in bioinformatics, and we believe there is a much greater potential for their applications in bioinformatics in the future.

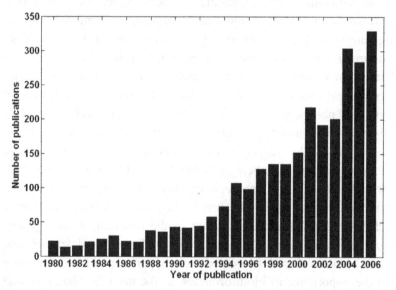

Figure 0.1. Number of publications containing the word "fuzzy" in their PubMed records since 1980.

Many biological systems and objects are intrinsically fuzzy as their properties and behaviors contain randomness or uncertainty. In addition, it has been shown that exact or optimal methods have significant limitations in many bioinformatics problems. Fuzzy set theory and fuzzy logic are ideal to describe some biological systems/objects and provide good tools for many bioinformatics problems. The applications of fuzzy concepts and approaches have been growing at an exponential rate. Figure 0.1 illustrates the number of publications that contain the word "fuzzy" in their titles or abstracts in PubMed, a literature database mainly for biomedical research. Currently the number is increasing at a rate of about one publication per day. While a number of books have been published covering applications of other computational intelligence techniques in bioinformatics, no book addresses the applications of fuzzy set theory and fuzzy logic in bioinformatics. As researchers in this area and educators at a university, we feel that there is an urgent need for a comprehensive and systematic book covering this topic. Hence, this is our motivation in writing this book.

We developed the text in a way that is useful to a broad readership, including students, postdoctoral fellows, and senior investigators moving into the field, as well as professional practitioners/bioinformatics experts. We expect that the book can be used as a textbook for upper undergraduate-level or graduate-level bioinformatics courses. Bioinformatics applications using fuzzy set theory or fuzzy logic often require good understanding of the biological background and the computational algorithms. In our case, no prerequisite in biology is needed, and only college-level calculus is required, for reading this book. In other words, a dedicated reader with a college degree in computational, biological or physical science should be able to follow the book without much difficulty. To facilitate learning and to maximize the benefit of the book, we provide a comprehensive introduction in fuzzy set theory and an appendix in basic biological concepts. We also wish to promote more research in applying fuzzy approaches in bioinformatics through this book, especially to provide an informative source for beginners entering bioinformatics as young students or as experienced researchers coming from other disciplines.

In this book, we discuss why and how fuzzy concepts and methods can play an important role in studying biological problems. We have designed the chapters to comprehensively address several important bioinformatics topics using fuzzy concepts and approaches. In addition, chapters have been connected seamlessly through a systematic design of the overall structure of the book. We start with an introduction to bioinformatics and then introduce fundamentals of fuzzy set theory and fuzzy logic. We focus on three examples (measurement of ontological similarity, protein structure prediction/analysis, and microarray data analysis). We also review other bioinformatics applications using fuzzy techniques. Finally we summarize and provide a future outlook. Furthermore we provide two appendices, one on fundamental biological concepts and one on online resources related to the book.

Chapter 1 (Introduction to Bioinformatics) discusses the scope of bioinformatics, including biological sequence analysis, protein structure analysis and prediction, gene expression data analysis, computational proteomics, gene ontology and biological pathway prediction. We will illustrate what the challenges in the fields are and why fuzzy logic can help.

Chapter 2 (Introduction to Fuzzy Set Theory and Fuzzy Logic) introduces fuzzy set theory and fuzzy logic. We will review the history of the field (together with types of successful applications). We will explain the key concepts and major methods, including fuzzy memberships, fuzzy clustering, fuzzy inference, etc. (tailored to potential bioinformatics applications).

Chapter 3 (Fuzzy Similarities in Ontologies) reviews some of the measures that can be used to compute the similarity between gene products annotated with terms from an ontology. We will introduce new fuzzy measures for computing ontological similarity between genes that avoid the problems of the traditional measures and, in addition, can account for information uncertainty. We will present several applications of the fuzzy similarity measures such as gene clustering and gene function summarization using the Gene Ontology terms. At the end of the chapter, we will present the application of the ontological similarity to computational intelligence algorithms such as fuzzy rule systems.

Chapter 4 (Fuzzy Logic in Structural Bioinformatics) introduces application of fuzzy logic in protein secondary structure prediction, protein solvent accessibility prediction, and protein structure comparison/classification. We will show our computational results and describe related computational tools.

Chapter 5 (Application of Fuzzy Logic in Microarray Data Analyses) provides a review of several microarray processing algorithms for gene selection and patient classification. We will then describe several clustering algorithms such as fuzzy c-means, relational fuzzy c-means and fuzzy co-clustering, and their use for gene selection.

Chapter 6 (Other Applications) reviews other types of bioinformatics applications using fuzzy set theory and fuzzy logic in the literature, including biological sequence motif identification, protein sequence alignment, protein subcellular localization prediction, 3D protein structure comparison, and computational proteomics.

Chapter 7 (Summary and Outlook) summarizes the whole book. We will discuss the advantages and limitations of using fuzzy set theory and fuzzy logic in bioinformatics. We will also provide an outlook of future applications and directions in using the fuzzy concept in molecular biology. Further related readings will be suggested.

Appendix I (Fundamental Biological Concepts) introduces some fundamental biological concepts for readers without a biological background. We will cover major biological subjects discussed in the book.

Appendix II (Online Resources) describes some of the free online resources, including tools, databases, and tutorials related to molecular biology, bioinformatics, and fuzzy set theory.

During the writing of this book, we have received help and support from our friends, colleagues, and families, to whom we wish to take this opportunity to express our deep gratitude and appreciation. First we would like to thank Imperial College Press, who contacted us to start this book project. During the writing of this book, Ms. Lenore Betts and Ms. Katie Lydon, editors at Imperial College Press, answered many of our questions and we are grateful their help. We like to thank Gerald L. Arthur, Tim Havens, Tran Hong Nha Nguyen, Yangjiong Su, Anders Wallqvist, and Jingfen Zhang for critically reviewing the drafts of the book and providing many helpful suggestions. We also want to thank our families for their constant support and encouragement over about a year of intensive writing.

Dong Xu
James Keller
Mihail Popescu
Rajkumar Bondugula

Contents

Chapter 1

Introduction to Bioinformatics

1.1 What Is Bioinformatics

As we enter the information age, we witness the impact of computers and computation in almost every corner of our lives. Many people in the world retrieve and broadcast information through the Internet. The weather forecast is made through extensive computation on supercomputers. Stocks are traded electronically. Airplanes are designed completely on computers before the first component is ever manufactured. We also witness substantial impact of computers and computation on biological and medical research, and this impact led to the birth of bioinformatics.

Although bioinformatics is a popular term in science and technology, there is no consensus for its definition. As a new field, its precise definition will take many years to finalize. A current semi-official definition for bioinformatics by the US National Institutes of Health (NIH) is "Research, development, or application of computational tools and approaches for expanding the use of biological, medical, behavioral or health data, including those to acquire, represent, describe, store, analyze, or visualize such data" (http://www.bisti.nih.gov/). A related field, computational biology, is defined by NIH as "the development and application of data-analytical and theoretical methods, mathematical modeling and computational simulation techniques to the study of biological, behavioral, and social systems". From these definitions, bioinformatics is focused on technology (engineering) for developing tools and infrastructure, while computational biology is more about science (biology) to generate hypotheses in understanding nature.

Although the distinction between bioinformatics and computational biology is made by NIH and others, there is no doubt that the two fields are tightly coupled. Hence, the terms bioinformatics and computational biology are sometimes used interchangeably. For example, the definition of bioinformatics by Luscombe *et al.* [2001] includes some scope of computational biology specified by NIH, but restricts itself to the biomolecular aspect: "bioinformatics is conceptualizing biology in terms of macromolecules (in the sense of physical-chemistry) and then applying "informatics" techniques (derived from disciplines such as applied math, computer science, and statistics) to understand and organize the information associated with these molecules, on a large-scale."

Bioinformatics is deeply rooted in three traditional disciplines, i.e., biology, computer science, and statistics. Both biology and computer science often claim bioinformatics as a sub-discipline. Furthermore, bioinformatics has strong ties to physics, biophysics, mathematics, chemistry, and engineering. On the other hand, bioinformatics is becoming an independent discipline by itself, with its own theoretical foundations, analytical approaches, and computational techniques. This emergence is similar to biophysics, which evolved from an interdisciplinary field between biology and physics to an integral science.

1.2 A Brief History of Bioinformatics

Although bioinformatics is a new term developed in the early 1990s, bioinformatics research started before 1970. Over the past four decades, bioinformatics emerged gradually from a hardly noticeable area to a mainstream discipline in science. You can find a comprehensive historical perspective of bioinformatics in the review by Ouzounis and Valencia [2003]. Here, we highlight some major milestones that define today's bioinformatics. If you are unfamiliar with some of the biological terms, you can find related materials in Appendices I and II.

In the 1960s, a number of key contributions in investigating biomolecular evolution paved the way for applying computers in studying biological sequences. Zuckerkandl and Pauling [1965] pioneered the use of biological sequences in evolutionary studies, which laid the theoretical foundation for computational studies of evolutionary patterns in genes and proteins. Fitch and Margoliash [1967] developed computational methods to build a tree structure (called "phylogenetic tree") from gene sequences for understanding gene evolution. Margaret Dayhoff and her coworkers developed a scoring method (called a "mutation matrix") for comparing protein sequences, and created computerized protein sequence databases for biomolecular evolution [Dayhoff *et al.,* 1965]. Because of her contribution, Dayhoff is regarded as a founder of the field of bioinformatics.

In the 1970s, a series of theoretical and computational studies opened new doors for bioinformatics research in diverse biological problems. Needleman and Wunsch [1970] published the first efficient algorithm for comparing two biological sequences based on dynamic programming. Lee and Richards [1971] provided a method for computing the geometry of protein three-dimensional structure. Chou and Fasman [1974] proposed a method for predicting protein secondary structures from a protein sequence. A few laboratories started simulation of protein dynamics and protein folding processes [Levitt and Warshel, 1975; Tanaka and Scheraga, 1975; Karplus and Weaver, 1976; Hagler and Honig, 1978]. Furthermore, RNA structure predictions emerged [Tinoco *et al.,* 1971; Waterman and Smith, 1978].

In the 1980s, various bioinformatics algorithms were significantly improved and bioinformatics tools became more sophisticated. In 1981, the Smith-Waterman algorithm for aligning two biological sequences was published [Smith and Waterman, 1981]. Although this algorithm is based on the one by Needleman and Wunsch [1970], the improvement allowed a comparison between parts of one sequence and parts of another sequence (which is called "local alignment"). This paved the way for large-scale sequence comparison and search. Because of this development and other contributions, Michael Waterman is regarded as another founder of bioinformatics. FASTA [Lipman and Pearson, 1985] was an early program for fast sequence similarity search in a database.

Feng and Doolittle [1987] developed a successful method to compare a group of sequences simultaneously (which is termed as "multiple-sequence alignment"). A number of systematic approaches for building phylogenetic trees were published, among which PHYLIP [Felsenstein, 1989] became a popular package. Kuntz *et al.* [1982] pioneered a method for predicting protein-ligand docking conformation. Computational methods for predicting genes from a DNA sequence were proposed [Shepherd, 1981; Fickett, 1982; Staden and McLachlan, 1982]. With these developments, the importance of bioinformatics research was recognized. Particularly, in 1988, the National Center for Biotechnology Information (NCBI) in the US was created to handle various bioinformatics issues from data distribution to data analysis.

The golden age of bioinformatics started in 1990s. This boom was mainly due to the Human Genome Project, which officially started in 1990. The goal of this project was to determine the sequence of the entire human genome. Genomic sequencing has opened a new avenue to study biological systems on large scales, setting the stage for generating many other high-throughput data. The new techniques for studying biology in large scale raised various new challenges for bioinformatics. Phil Green and his colleagues addressed the computational problem of identifying nucleotides from image data of a sequencer, a process referred to as "base calling" [Ewing *et al.*, 1998]. A widely used method for genome sequencing is the "shotgun" approach, where bioinformatics is required to assemble short, overlapping pieces of DNA sequences into a long, coherent sequence. Green [2002] and Myers [1995] developed methods for solving this problem, which was a major contribution to the Human Genome Project.

In 1990, the exponential growth of biomolecular data clearly showed the need for interpreting, managing and mining these data. Various bioinformatics databases, such as GenBank (http://www.ncbi.nlm.nih.gov/Genbank), a database for biological sequences, became essential to biomedical research. Many bioinformatics algorithms led to sophisticated computer packages, with user-friendly interfaces. Meanwhile, computers became faster and cheaper and the Internet provided a major platform for accessing bioinformatics tools and databases. Many experimental biologists started to use various bioinformatics packages, especially through Web interfaces [Xu *et al.*, 2000; Rhee *et al.*, 2006]. A number of popular software packages and servers developed in the 1990s are widely used, as indicated by their large numbers of citations (see Table 1.1). The sequence comparison tool BLAST [Altschul *et al.*, 1990] became a household name to biologists. It is also the most popular tool among all the computational tools that have ever been developed since the birth of the computer, with its defining paper as the most cited reference in the scientific history. The widespread use of bioinformatics applications has had an enormous impact on research in biology and medicine.

Table 1.1. Popular Bioinformatics Packages

Name	Functionality	URL	Reference	Citations
BLAST	Pairwise sequence alignment	http://www.ncbi.nlm.nih.gov/BLAST/	Altschul *et al.*, 1990	20,495
CLUSTAL-W	Multiple sequence alignment	http://www.ebi.ac.uk/clustalw/	Thompson *et al.*, 1994	18,837
SignalP	Signal peptide prediction	http://www.cbs.dtu.dk/services/SignalP/	Nielsen *et al.*, 1997	3002
DALI	Protein structure comparison	http://www.ebi.ac.uk/dali/	Holm and Sander, 1993	g
MODELLER	Protein tertiary structure prediction	http://www.salilab.org/modeller/	Sali and Blundell, 1993	1817
PHD	Protein secondary structure prediction	http://www.predictprotein.org/	Rost and Sander, 1993	1795
SEQUEST	Protein identification using mass-spec data	http://fields.scripps.edu/sequest	Eng *et al.*, 1994	1324
MFOLD	RNA secondary structure prediction	http://www.bioinfo.rpi.edu/applications/mfold/	Mathews *et al.*, 1999	1228
PHRED	DNA sequencing	http://www.phrap.org/	Ewing *et al.*, 1998	1162
GENESCAN	Gene identification in DNA	http://genes.mit.edu/GENSCAN.html	Burge and Karlin, 1997	1139

Note: The number of journal citations was based on the "ISI Web of Knowledge" (http://nadc.isiknowledge.com) on August 4, 2006.

Coming into the new millennium, bioinformatics became a very active research field as modern biology quickly evolves. The availability of genomic sequences enabled a number of new high-throughput measurement technologies, which expanded genome-scale studies from sequence-level information to higher-level functions. For example, microarrays are powerful tools for systematic measurement of large-scale gene-expression data under varying experimental conditions or over a time course. Various experimental methods can generate different types of protein-protein interaction information [Chen and Xu, 2003]. Each new experimental technique for large-scale biomolecular measurement often requires new development in data interpretation and analysis. In microarrays, extensive studies have been conducted in image processing, statistical analysis, and clustering [Speed, 2003]. In recent years, new sequencing techniques [Service, 2006], such as the 454 sequencer [Margulies *et al.*, 2005], were developed to reduce the cost of sequencing. The bioinformatics challenges in these sequencing technologies are numerous in terms of experimental design, data interpretation, and data integration.

In recent years, systems biology emerged as a field which integrates experimental, theoretical, and computational techniques to study biological organisms at multiple levels as a system instead of individual components [Alon, 2006]. A systems approach brings renewed hope for solving some long-standing biomedical problems, especially various complicated diseases such as cancer and diabetes. Bioinformatics is a key component in systems biology, bringing heterogeneous data together for analysis, modeling and design [Kriete and Eils, 2005]. For example, bioinformatics can be used to predict a biomolecular network as a large-scale system [Palsson, 2006]. In addition, it also helps to fuse and integrate a wide spectrum of high-throughput data, including biological sequences, gene expression levels, protein interactions, small RNA regulation [Washietl *et al,.* 2005], epigenomics data [Model *et al.,* 2001], and metabolomic data [Steuer *et al.,* 2003].

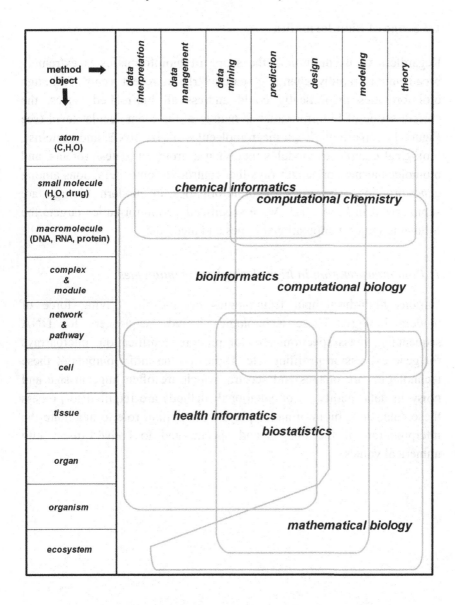

method ➡ object ⬇	*data interpretation*	*data management*	*data mining*	*prediction*	*design*	*modeling*	*theory*
atom (C,H,O)							
small molecule (H_2O, drug)							
macromolecule (DNA, RNA, protein)							
complex & module							
network & pathway							
cell							
tissue							
organ							
organism							
ecosystem							

chemical informatics

computational chemistry

bioinformatics

computational biology

health informatics

biostatistics

mathematical biology

Figure 1.1 Our view of the scope of bioinformatics and related areas in a matrix of biological objects and computational approaches.

1.3 Scope of Bioinformatics

Regardless of its definition, the scope of bioinformatics is extremely broad and is rapidly changing, particularly in recent years. Although bioinformatics theoretically could address all bio-related issues, the current scope of bioinformatics is mainly at the biomolecular level (see Figure 1.1), particularly on macromolecules (DNA, RNA, and proteins), biological complexes/modules involving a group of genes/proteins, and biomolecular networks/pathways that control various interactions among genes/proteins. The roles of bioinformatics in modern biology are summarized in Figure 1.2. More specifically, bioinformatics targets the following major computational issues and methods:

1) Data interpretation in high-throughput technologies

Various high-throughput technologies became the driving force of modern biology. These technologies include sequencers for DNA sequencing, mass spectrometers for protein identification, microarrays for gene expression profiling, etc. Typically, the initial outputs of these technologies are images and spectra, which are often huge in size and noisy in data quality. Computational methods are required to process these data; thus, bioinformatics plays an important role to automate the interpretation of the images and spectra and to convert them into numerical values.

2) Data management and computational infrastructure

Given the size and complexity of biological data, creation and maintenance of databases of biological information are essential to modern biology. Biological sequences and their annotations comprise the majority of such databases, while many other types of databases for microarray gene expression, protein structures, etc. are expanding quickly. Bioinformatics handles the design of these databases for data storage, update, and retrieval. In many cases, a Web interface is provided for data access, together with some back-end engines for data analyses (see Appendix II for examples). Sometimes graphical tools or plug-ins are provided for data visualization. Furthermore, some biological databases may connect to experimental instruments for real-time data collection using a tracking system, such as a Laboratory Information Management Systems (LIMS) [Paszko and Turner, 2001].

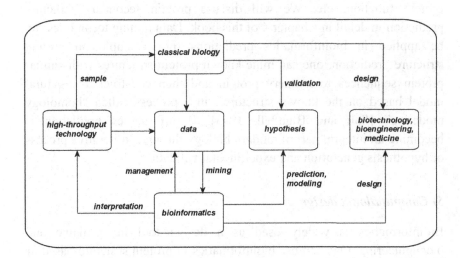

Figure 1.2 Roles of bioinformatics in modern biology.

3) Discovery from data mining

A demanding task for bioinformatics is to extract useful biological information and patterns from noisy data produced by high-throughput technologies. For example, one can compare sequences of multiple genomes to identify interesting evolutionary patterns. Analyzing microarray data can lead to the discovery of the genes that are associated with a particular disease. Mining biomedical literature can lead to automated identification of possible gene-gene associations. We will address microarray data mining extensively in Chapter 5 of this book.

4) Prediction

Bioinformatics is often used to predict biological information. In particular, from a protein sequence alone, one can predict protein secondary structure, protein localization (in a compartment of a cell), protein function, etc. We will discuss protein secondary structure prediction in detail in Chapter 4 of this book. Data mining techniques can be applied in bioinformatics predictions. For example, in protein structure prediction, one can mine known protein structures with similar protein sequences to a query protein and then construct a structural model based on the known structures, in a process called "homology modeling" [Sali and Blundell, 1993]. Bioinformatics prediction is becoming an integral part of modern biology through an iterative process of hypothesis generation and experimental validation.

5) Computational design

Bioinformatics is widely used as a design tool in medicine and bioengineering. One can use bioinformatics in protein structure-based or gene-based drug discovery and development. It can help design a delivery using a certain combination of drugs and at a certain schedule to achieve maximum performance (e.g., for AIDS treatment). It can also suggest mutations of a gene for achieving certain biological properties in a genetically modified species. For example, it is possible to suggest a mutation of a gene to achieve drought resistance in soybeans.

6) Modeling

Modeling of biological systems and processes often adds value to the available biological data. A well established area in bioinformatics is protein structure modeling [Xu *et al.*, 2006]. One can model various aspects of a protein structure, including geometry, energetics, and dynamics. For example, a useful modeling technique [Nicholls *et al.*, 1991] calculates and visualizes the electrostatic field of a protein structure. One can also model a neural system using differential equations, with parameters fitting some experimental data. In many cases, these parameters cannot be measured directly. Modeling can be used to interpret experimental results and to generate new hypotheses. In recent years, an entire cell has been modeled. For example, E-Cell (http://www.e-cell.org/) attempts to model and reconstruct biological phenomena computationally and perform whole cell simulations.

As bioinformatics expands its scope, a number of areas emerge as sub-disciplines. Each of the sub-disciplines has its own special methods and techniques. Structural bioinformatics focuses on the computational analysis and prediction of macromolecular structure (especially protein structure). Computational proteomics handles management and analysis of proteomics data for protein identification and protein interaction determination. Computational systems biology addresses algorithm and application development for systems biology. On the application side, sub-disciplines focus on the application of bioinformatics in different biological subjects. For example, immunoinformatics models immunological components for better understanding immune functions. Pharmacoinformatics deals with drug discovery using bioinformatics approaches. Agroinformatics (agricultural informatics) specializes in the bioinformatics that deals with plants and domestic animals.

In addition to expansion in various sub-disciplines, there are overarching issues for bioinformatics. One of them is bioinformatics standards. As almost all the analyses in bioinformatics are large scale, automated processing without extensive manual interruption is essential. For different tools and databases to communicate with each other, some standards need to be established. One of the efforts is ontology, which is a set of controlled vocabularies. We will have a thorough discussion of ontologies in Chapter 3. An infrastructure to facilitate interactions between databases and servers is the semantic web, which creates a universal mechanism for information exchange and reuse in a machine-interpretable way across application, organization, and community boundaries [Neumann, 2005].

Bioinformatics has a number of related fields (in addition to computational biology), as illustrated in Figure 1.1. At the small molecule level, cheminformatics applies information technology in identification and optimization of drug leads, while computational chemistry employs quantitative methods for calculating molecular properties or simulating molecular behavior. At the macroscopic scale, health informatics (or medical informatics) addresses computational development for improving communication, understanding and management of medical information and practice; biostatistics applies statistical techniques in health-related fields, such as medicine, biology, and public health; and mathematical biology (or biomathematics) uses theoretical and numerical methods and tools to model biological systems and processes. The areas shown in Figure 1.1 overlap with each other and their scopes have been historically defined, although different researchers have different views. Among all the areas, a unique hallmark of bioinformatics is its emphasis on the development of computational tools and infrastructures driven by the need of users instead of the developers.

1.4 Major Challenges in Bioinformatics

As massive biological data have become a fundamentally important resource during discovery of new biological knowledge, a key task for bioinformatics is to identify meaningful information (or statistically significant patterns) from data and correlate such information with biological knowledge. However, such a task is highly challenging in many cases: (1) the data size is large with high dimensionality, with a complexity much higher than those typically handled by traditional computational sciences; (2) the information-rich data are heterogeneous in nature, noisy, and incomplete, as well as containing misleading outliers; and (3) biological systems, due to adaptability, evolution, redundancy, robustness, and emergence, are extremely complex. Many biological data are generated by biological processes which are not well understood. Interpretation of such data requires discovery of convoluted relationships hidden in the data. Due to these challenges, the accuracy of prediction or the information mined from a database is often not satisfactory. It is clear that there is much room for further improvement and development, which require novel theoretical frameworks and computational techniques.

Other than the technical challenges, human factors are also important. Given the scope of bioinformatics, it is unlikely for a single person to have deep understanding in relevant fields of computer science, biology, and statistics. Inevitably a researcher may not have a complete view or knowledge to solve a particular problem. In most cases collaborations are needed. However, overcoming "language" barriers among researchers from different backgrounds is often demanding. Currently, a majority of experimental biologists are not familiar with concepts, methods and tools available or emerging in bioinformatics. Computational researchers often do not understand biology in depth. More communication among different disciplines is essential for bioinformatics research.

1.5 Bioinformatics and Computer Science

The challenges in bioinformatics have resulted in a wide range of studies from computer sciences. Almost all available computer science techniques have been applied in bioinformatics. The following are some most notable applications of computational and statistical methods in bioinformatics:

1) Dynamic programming
2) Neural networks
3) Hidden Markov Models
4) Hypothesis test
5) Bayesian statistics
6) Clustering
7) Sampling search (Gibbs, Monte Carlo, etc)
8) Maximum likelihood methods
9) Information theory
10) Support Vector Machines

Fuzzy set theory has been used in bioinformatics, but to a much less extent than any of the methods above. We believe there is much higher potential for fuzzy set theory in bioinformatics, and hence, the focus of this book.

Not only does computer science provide techniques for bioinformatics, bioinformatics is also a new driver of computer science. Better hardware (supercomputers) is often demanded by bioinformatics applications. New data representation and new algorithm development fuel active research in computer science. Bioinformatics may also inspire new theoretical frameworks for computer science. Traditionally, a number of computational techniques came from biological concepts, such as neural networks, genetic algorithms, automata, and fuzzy set theory. In recent years, DNA computing is being developed to use DNA and biochemical reactions, instead of the traditional silicon-based computer chips to solve computational problems [Adleman, 2004]. Ant colony optimization mimics the behavior of ants in finding paths from the colony to food and uses a probabilistic technique for solving computational problems [Dorigo and Stützle, 2004]. Meanwhile particle swarm optimization [Eberhurt and Kennedy, 1995] imitates social communication (say, among insects) to produce cooperation in groups of potential solutions in hunting for very good answers to highly complex problems. More computer science techniques will be developed with the research of bioinformatics.

If the reader wishes to know more about bioinformatics, we suggest some related books for reference [Jiang *et al.*, 2002; Claverie, 2003; Jones and Pevzner, 2004; Lesk, 2005] and review articles [Luscombe *et al.*, 2001; Ouzounis and Valencia, 2003; Kanehisa and Bork, 2003; Bonetta, 2004; Rhee *et al.*, 2006], as well as the Internet resources listed in Appendix II.

Chapter 2

Introduction to Fuzzy Set Theory
and Fuzzy Logic

2.1 Where Does Fuzzy Logic Fit in Computational Science?

There are many computational models and approaches that can be applied to bioinformatics problems. These fields of study can be aggregated under the general heading of computational science. Our view of one possible categorization of the various disciplines within computational science is shown in Figure 2.1. We believe that the ever expanding group of disciplines covered by computational intelligence (neural networks, fuzzy systems, evolutionary computation, swarm intelligence, autonomous mental development, etc.) occupies a central place in that taxonomy. These are all paradigms that take inspiration from biological properties and animal or human characteristics and intelligence to create computing models. While all approaches from computational intelligence have a significant role in bioinformatics, we argue that the very nature of biology, and in particular, producing automated reasoning algorithms about biology, cries out for the use of fuzzy set theory and fuzzy logic. This book is our attempt to demonstrate the richness of design that can be realized through fuzzy set models in bioinformatics.

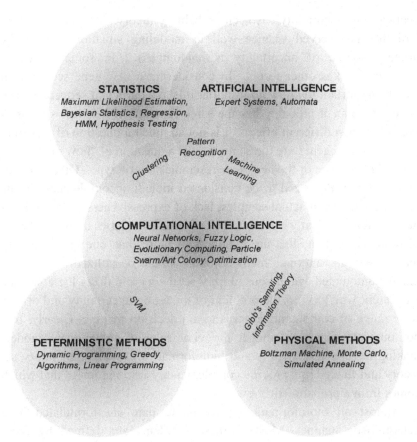

Figure 2.1 Our view of the central role of computational intelligence within the general field of computational science.

2.2 Why Do We Need to Use Fuzziness in Biology?

Throughout the history of science, there has always been a need to model and manage uncertainty in the physical world. This is particularly true in biology in general, and more recently, bioinformatics. The variability exhibited in nature in studying the genome and its relationship with phenotypical behavior require theoretical and computational models to be flexible enough to capture the essential aspects without "seeing" every deviation as something completely new. The historical framework for dealing with uncertainty has been probability theory. This is a powerful tool that has served science well in modeling situations where the primary source of uncertainty is randomness. In some instances, uncertainty takes other forms. In considering a new gene sequence, it may be important to know how similar it is to a particular sequence – if it is very similar, it probably has the same function, if it is less similar, it may produce a different effect. It is not so much a question of "whether or not" the two genes are the same, as it is a question of "how much" this particular instance of the new gene resembles a prototype. Other sources of uncertainty that need to be considered include incompleteness in the data extracted from actual samples, lack of expressiveness or faithfulness of some features that we extract, lack of clear boundaries between classes of proteins, proteins that are members of more than one class, etc. In these situations, alternate methodologies should be utilized to aid us in making automated evaluations. Fuzzy set theory and fuzzy logic provide a different way to view the problem of modeling uncertainty and offer a wide range of computational tools to aid decision making. Clearly, it is not our intention to diminish the vital role of probabilistic models making in science generally and bioinformatics in particular. Fuzzy set theory and fuzzy logic provide complementary information to that which comes from a probabilistic view.

Almost all bioinformatics problems to date are formulated in a deterministic manner. Most of these problems are defined by fixed objective functions and solved through optimization. Many dynamic processes, such as gene expression regulation, are also modeled using differential equations with deterministic behavior. However, there are at least three situations in which fuzziness should be considered, i.e., intrinsic fuzziness in biological systems, multiple roles of a biological object, and fuzzy descriptions of biological phenomena.

In recent years, there is an increasing awareness of the fuzzy aspects of biological systems, which is sometimes referred to as a paradigm shift for "new biology". An expanding body of evidence has been found that many processes in biological systems are intrinsically fuzzy rather than deterministic. Numerous examples have demonstrated that fuzzy effects are physiologically and evolutionarily important in the development and function of living organisms. For example, it was found that the immune repertoire, as a consequence of central tolerance, is able to recognize both self and non-self antigens in a fuzzy manner. In this case, the key players of the immune system, T cells and antibodies, can recognize given self or "foreign-reactive cells" (non-self) to a certain degree, although deterministically. Such a fuzzy feature of the immune system may shed light on mechanisms of autoimmune diseases, such as systemic lupus erythematosus. Fuzziness can be achieved by random fluctuations. For example, random fluctuations are intrinsically important for balancing fidelity and diversity in eukaryotic gene expression and may produce variability in cellular behavior. The stochastic dynamics can also provide additional functional modalities on the enzymatic futile cycle mechanism that include stochastic amplification and signaling. This stochastic/fuzzy behavior may offer a novel type of control mechanism in pathways that contain these cycles.

A biological object may have multiple roles, resulting in fuzzy memberships in each role. One gene may be involved in different functions or pathways. Beta-catenin is a multifunction protein, playing important roles in both cell-cell adhesion and intracellular signaling. In practice, when we cluster genes using biological data (e.g., microarray gene expression data), fuzzy memberships for a gene may be useful to serve as descriptors for that gene in the context of fuzzy clustering with a large set of genes. In this case, one gene can be present in two or more clusters simultaneously, with partial or full membership in each cluster.

Another type of fuzziness in biomedical research results from the fuzzy description of biological terms. Our descriptions of many biological concepts often have difficulty fitting into a deterministic (crisp) explanation. As a result, our knowledge, concepts, and representations of biological terms may also be fuzzy, and fuzzy set theory is useful to describe these terms. For example, the species definitions for microbes can be fuzzy due to recombination of the genetic materials across species [Hanage *et al.*, 2005]. The concept of "protein function" is sometimes fuzzy because it is often based on whimsical terms or contradictory nomenclature [Jansen and Gerstein, 2004]. This currently presents a challenge for functional genomics. In addition, descriptions for similarity and typicality can be fuzzy. For example, how much do two proteins resemble each other, what properties do they (partially) share, how close is a given protein to the prototypical sequence of a protein family, etc. Such fuzziness could result from the limitations of classifications, natural language, or poor understanding of the underlying mechanism. Tolerance of fuzziness allows us to explore these biological concepts effectively.

Fuzzy set theory in general and fuzzy logic specifically are natural ways to model ambiguous events that occur in human-like reasoning. People have no trouble operating with phrases such as "large risk factor", "somewhat likely to be involved in cancer", "significantly hyper methylated", etc. As will be seen, rules containing such ambiguous clauses can be successfully handled in a fuzzy logic system.

The beauty (and also a danger, if we are not careful) of fuzzy set theory is that it offers a multitude of calculi for the fusion of partial support for a hypothesis under investigation, that is, flexible mechanisms to increase or decrease confidence in a decision as evidence unfolds. In his seminal text on computer vision, [Marr, 1982], David Marr stated two principles to be followed in the design of intelligent (vision) algorithms. The first is called the Principle of Least Commitment (PLC). He states it simply as "Don't do something that later must be undone". Hence, in a complex computing scenario, one where there are many decision making steps, avoid making deterministic decisions for as long as is possible. It is very difficult, perhaps impossible, to recover from a wrong crisp decision early on. Keep your options open until the situation demands a final answer. While Marr was interested in computer vision, the PLC certainly applies to bioinformatics applications. As a simple illustration, consider the problem of classifying new genes by their DNA sequence. Sequencing machines actually produce memberships (or probabilities) for all four DNA bases (A, C, T, G) at each position. Most of the time, the nucleic acid with highest membership is chosen and, from that point on, the sequence is treated as if it were deterministic. Hence, potentially valuable information is discarded and is unavailable in subsequent processing. If this now deterministic sequence is matched to a database and, say, only the top match is considered, additional information is lost. Clearly, the concept of assigning and maintaining degrees of membership (perhaps confidence in competing hypotheses) or more general linguistic labels in fuzzy set theory support the PLC for complex decision making applications as in bioinformatics.

The second principle of Marr is called the Principle of Graceful Degradation (PGD). By this he meant that algorithms should delivery a partial (reasonable) answer as input degrades. In other words, intelligent algorithms should encompass a degree of robustness and continuity. Here also, techniques that utilize membership degrees or other fuzzy constructs in the calculation of their response to input conditions have the potential to degrade much more gracefully than their crisp counterparts. Consider, for example, a simple cancer screening task that relies on a single test value, say the amount of expression of a particular gene in a microarray assay (not overly realistic, but for illustrative purposes). If the outcome is binary (cancer/non-cancer) and the algorithm is crisp, say based on a threshold, then a slight amount of noise in the test score can actually flip the screening result from non-cancer to cancer or vice versa. If however the output is a membership in the cancer diagnosis modeled as a trapezoid function around the threshold value [Klir and Yuan, 1995], such small perturbations only change the degree of risk in a correspondingly small amount. Figure 2.2 shows a typical trapezoidal curve, along with other standard fuzzy membership functions. This overly simple example illustrates the point that fuzzy models embrace the concept of the PGD.

2.3 Brief History of the Field

Concepts of vagueness and fuzziness have been contemplated in mathematics and science for quite awhile. For example, in 1923 Bertrand Russell stated "All traditional logic habitually assumes that precise symbols are being employed. It is therefore not applicable to this terrestrial life, but only to an imagined celestial existence." [Russell, 1923]. Like Russell, the philosopher Max Black was concerned with vagueness and imprecision in language, and the effect of these concepts on logic [Black, 1937]. In fact, he believed that all terms whose application involves using our senses are vague. Black, in 1937, actually came up with the concept that we now associate with membership functions. He even conducted a cognitive psychological experiment with a group of people that effectively constructed membership functions exemplifying vagueness of certain words. However, most people attribute the beginning of fuzzy set theory to Lotfi A. Zadeh's 1965 paper [Zadeh, 1965] that developed this topic in its current form. An excellent treatment of the history of fuzzy sets and fuzzy logic can be found in [Seising, 2005]. The journal Fuzzy Sets and Systems published a "40th Anniversary of Fuzzy Sets" in December 2005 that contains 14 position papers covering various aspects of the role and future prospects of fuzzy sets [Dubois, 2005].

The mathematical basis for formal fuzzy logic can be found in infinite-valued logics, first studied by the Polish logician Jan Lukasiewicz in the 1920s (see [Borkowski, 1970]). Lukasiewicz constructed a series of multi-valued logical systems, generalizing from small finite numbers of truth-values to those containing infinite sets of truth values. His work and calculation formulae are ingrained in modern fuzzy set theory and fuzzy logic, the genesis of which is credited to Zadeh in his seminal three part treatise on the theory and applications of linguistic variables [Zadeh, 1975a; Zadeh 1975b; Zadeh 1976].

Perhaps the biggest boost to the visibility and perceived utility of fuzzy set theory came from the application of rule-based fuzzy systems to problems in control [Mamdani and Assilian, 1975; Mamdani, 1977; Takagi and Sugeno, 1985; Sugeno, 1985, Verbruggen and Babuska, 1999, Passino and Yurkovich, 1998]. In what has become commonplace now, sets of linguistically described rules were created and inserted into a variety of non-linear control systems. The ease of design and the smoothness of the control surface from only a handful of rules made fuzzy controllers very popular in a variety of products from the automotive industry, consumer electronics markets, etc. Fuzzy controllers are well suited for low-cost embedded systems.

While the big economic impact of fuzzy set theory and fuzzy logic centers on control, particularly in consumer electronics, there has been, and continues to be, much research and application of these technologies in pattern recognition, information fusion, data mining, and automated decision making [Bezdek *et al.*, 1999, Keller *et al.*, 1996]. There are national, multi-national, and international fuzzy systems professional societies around the globe whose purposes are to foster research, development and application of fuzzy set theory and fuzzy logic. Fuzzy systems are one of the core pillars of the IEEE Computational Intelligence Society.

An introduction to the key components of fuzzy set theory and fuzzy logic is now given with the view toward computational methods of use in bioinformatics. After a discussion of the general principles of fuzzy set theory, membership functions and fuzzy connective operators, we focus on those areas for which we present specific applications within bioinformatics: fuzzy logic rule based systems, fuzzy clustering, fuzzy classifiers, particularly, the Fuzzy K-Nearest Neighbor algorithm, fuzzy measures and the fuzzy integral. The reader is referred to [Klir and Yuan, 1995; Bezdek *et al.*, 1999] for more extensive development of the theory and selected applications.

2.4 Fuzzy Membership Functions and Operators

2.4.1 Membership functions

Traditional set theory is based on binary, or two-valued, logic. Given a "universe" set X, a subset A of X can be defined in several ways. Suppose that X is the set of integers. The subset of even positive integers can be specified by listing its members:

$$A = \{2,4,6,8,\cdots\}$$

or by providing defining properties

$$A = \{x \in X | x \text{ is an even positive integer}\}.$$

Alternately, we define a subset A by its characteristic function, which is also denoted by the set name, $A : X \rightarrow \{0,1\}$ from X into the binary set $\{0,1\}$ given by

$$A(x) = \begin{cases} 1 & \text{if } x \in A \\ 0 & \text{if } x \notin A. \end{cases}$$

Zadeh [1965] simply defined a fuzzy subset of X as a function $A: X \rightarrow [0,1]$, i.e., a characteristic function from X into the interval $[0,1]$. The value $A(x)$ is called the membership of the point x in the fuzzy set A or the degree to which the point x belongs to the set A. For example, the fuzzy subset of "big positive even integers" could be defined by

$$A(x) = \begin{cases} 1 - \dfrac{2}{x} & \text{if } x = 2,4,6,\cdots \\ 0 & \text{else.} \end{cases}$$

An example closer to the topic of this book is a membership function, A, that describes the similarity of a protein sequence S to that of a specific sequence called S' in a protein database. A typical measure for sequence similarity is the expectation value, *Eval*, returned from BLAST [Altschul *et al.*, 1997]. Assume the score for the alignment between S and S' under a scoring scheme is R. *Eval* indicates the number of different alignments with scores equivalent to or better than R that are expected to occur in a database search by chance. Although expectation value is a good descriptor for sequence search from a database, it does not reflect all the information of the protein similarity. For example, when two sequences are identical, the expectation value varies accordingly to the length, although biologically the two proteins are the same. Two different long sequences can have better expectation value than two identical short sequences. To address this issue, we can describe the similarity between any protein sequence and one target protein sequence from biological point of view using a fuzzy membership function as an alternative measure, e.g.,

$$A(S) = \begin{cases} 0 & \text{if } Eval > 0.1 \\ sim(S,S') & \text{else} \end{cases}$$

where *sim(S,S')* is the sequence identity of the alignment, i.e., the percentage of the identical amino acids in the alignment.

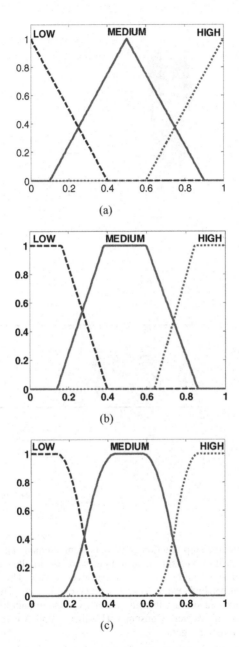

Figure 2.2 Examples of common fuzzy membership functions. (a) triangular, (b) trapezoidal, and (c) smooth quadratic functions.

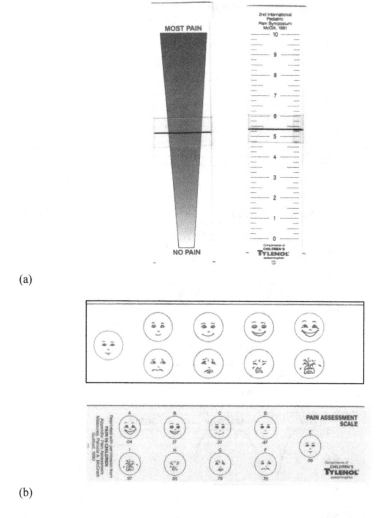

(a)

(b)

Figure 2.3 Practical membership function generators from a nursing application in having children assess their level of pain for medication dosage. In (a), the child slides a bar to the place on the visual scale that best represents her level of pain. The nurse can read off the analog "pain membership" value from the reverse side. (b) This is for younger children and demonstrates a similar, though discrete, version of pain membership. These data are the property of McNeil Consumer Healthcare and McNeil maintains sole publication and dissemination rights.

All fuzzy set theory is based on the concept of a membership function. Where do these membership functions come from? In many cases, they are defined as in the two examples above – common sense definitions that convey some linguistic expression. More generally, they come from expert knowledge directly or they can be derived from questionnaires, heuristics, etc. This is a human-centric view and is certainly open to debate. In many cases, the membership functions take on specific functional forms like triangular, trapezoidal, S-functions, pi-functions, sigmoids, and even Gaussians for convenience in representation and computation. Pi-functions and S-functions are constructed from quadratic functions "pieced together" to make smooth curves.   Figure 2.2 displays several common fuzzy membership functions. Alternately, membership functions (or the parameters of the specific equation forms) can be learned from training data, much as probability density functions are learned.   Some fuzzy clustering algorithms naturally produce membership functions as their output. A neural network, given proper input/output training data, also acts as a membership function for new input.

One of our favorite practical membership functions comes from the field of pediatric nursing. A child is asked to slide the bar to a position indicative of his or her level of pain in Figure 2.3(a). The color and width provide a guide. The nurse turns the instrument over to recover what is effectively a membership (after dividing by 10) in the fuzzy set "Pain". The goal is to provide sufficient pain medication without over dosing. This is a continuous membership. For very young children, see the discrete memberships as shown in Figure 2.3(b).

2.4.2 Basic fuzzy set operators

Once fuzzy subsets of a universal set X are defined, definitions for the complement of a set, the union of two sets and the intersection of two sets are required to actually generate a "set theory". In 1965, Zadeh proposed the following. Suppose $A : X \rightarrow [0,1]$ is a fuzzy subset of X. The complement A^c of A is defined by

$$A^c(x) = 1 - A(x).$$

Additionally, if $B : X \rightarrow [0,1]$ is another fuzzy subset of X, Zadeh defined

$$(A \cup B)(x) = \max\{A(x), B(x)\} = A(x) \vee B(x)$$

and

$$(A \cap B)(x) = \min\{A(x), B(x)\} = A(x) \wedge B(x).$$

Why did Zadeh define the operators in this manner? Quite simply, it was because these definitions revert back to the standard crisp definitions if the subsets are crisp. Hence, this forms a true extension of normal set theory. As can be found in the many textbooks on fuzzy set theory (see [Klir and Yuan, 1995; Pedrycz and Gomide, 1998] for example), all of the theorems of crisp set theory hold for this fuzzy set theory except two: the Law of Contradiction (LOC) and the Law of the Excluded Middle (LEM). The LOC states that the intersection of a set and its complement must be empty ($A \cap A^c = \phi$), while the LEM requires that the union of a set and its complement must be the whole universe set ($A \cup A^c = X$). Since crisp set theory is formally equivalent to the first order predicate logic, these two laws state that a proposition can not be both true and not true simultaneously, and that either a proposition or its negation (complement) must be true. While these statements seem reasonable, they give rise to a paradox within classical logic, commonly called Russell's paradox. A simple version goes something like this: Russell's barber has a sign that states "I shave everyone, and only those, who do not shave themselves". Then who shaves the barber? If he shaves himself, then he can not (shaves only those who do not shave themselves); but if he does not shave himself, then he must (shaves everyone who don't shave themselves). Such a dilemma! As pointed out earlier, some biological entities have multiple roles. Hence, it is impossible to put them only into one set; they may also naturally fit into the complement set. It is a mistake, for example, to put Beta-catenin only in the cell-cell adhesion subset since it also clearly belongs in the complement of the cell-cell adhesion subset (involved in intercellular signaling). So, perhaps it is not that unreasonable to disobey the LOC and the LEM.

Suppose that the membership function values for a particular element x in X are interpreted as the confidences that x possesses certain properties, e.g., $A(x)$ is the confidence that x is involved in cell-cell adhesion and $B(x)$ corresponds to the confidence that x functions in intercellular signaling. Then the original Zadeh definitions of complement, union, and intersection produce confidences related to the linguistic concepts of NOT, OR, and AND: $A^c(x)$ is the confidence that x is NOT involved in cell-cell adhesion; $(A \cup B)(x)$ gives the degree to which x is either involved in cell-cell adhesion OR x is involved in intercellular signaling; $(A \cap B)(x)$ computes the confidence that x is involved in cell-cell adhesion AND x is involved in intercellular signaling. Beta-catenin would have a non-zero membership in the intersection of A and A^c.

The good news and the bad news in fuzzy set theory is that there are infinite numbers of ways to define complement, union and intersection [Klir and Yuan, 1995; Dubois and Prade, 1985]. An alternate fuzzy set theory that is useful for fuzzy logic inference is generated by the operators

$$(A \cup_b B)(x) = 1 \wedge (A(x) + B(x)),$$

$$(A \cap_b B)(x) = 0 \vee (1 - (A(x) + B(x))),$$

called the bounded sum and bounded difference, along with the standard complement,

$$A^c(x) = 1 - A(x).$$

Each such extension of crisp set theory loses either LOC and LEM or two other properties (idempotency and distributivity) [Klir and Yuan, 1995]. For this choice, LOC and LEM are satisfied, while idempotency and distributivity are lost. Actually, there are infinite families of union, intersection and complement operators that are extremely useful in multicriteria decision making where partially supported criteria are to be combined in disjunctive (OR) and/or conjunctive (AND) manners to reach an overall evaluation of an alternative.

One such infinite family of connectives is due to Yager [Yager, 1980]. Here, complement, union and intersection are given by

$$A^c(x) = (1 - A(x)^w)^{1/w}, w \in (0, \infty),\qquad (2.1a)$$

$$(A \cup_w B)(x) = \min\{1, (A(x)^w + B(x)^w)^{\frac{1}{w}}\}, w \in (0, \infty), \text{ and } (2.1b)$$

$$(A \cap_w B)(x) = 1 - \min\{1, ((1 - A(x))^w + (1 - B(x))^w)^{\frac{1}{w}}\}, \quad (2.1c)$$
$$w \in (0, \infty).$$

For all choices of w, the value of the Yager union operator is greater than the standard union (max), while that for the intersection is less than the standard intersection (min). In other words, a Yager union operator is more optimistic than the maximum (in combining confidence), whereas each Yager intersection produces values that are more pessimistic than the minimum. The parameter w controls the degree of optimism or pessimism. In fact, the following limits hold:

$$\lim_{w \to 0} (A \cup_w B)(x) = A(x) \vee B(x) \text{ and}$$

$$\lim_{w \to 0} (A \cap_w B)(x) = A(x) \wedge B(x).$$

At the other end, i.e., the limits as $w \to \infty$, generate the drastic union and intersection, defined by

$$(A \cup_d B)(x) = \begin{cases} A(x) & \text{if } B(x) = 0 \\ B(x) & \text{if } A(x) = 0 \text{ and} \\ 1 & \text{else} \end{cases}$$

$$(A \cap_d B)(x) = \begin{cases} A(x) & \text{if } B(x) = 1 \\ B(x) & \text{if } A(x) = 1. \\ 0 & \text{else} \end{cases}$$

Besides their use in what is to come, fuzzy operators have been used extensively in multicriteria decision making [Bellman and Zadeh, 1970; Yager, 1988; Yager, 2004]. We end this section with an example of the use of fuzzy set operators in multicriteria decision making.

Example 2.1. As a particularly simplistic illustration, consider a decision tree, as in Figure 2.4, to assess cancer risk based on the following observations. Suppose we decide that cancer risk should be high if either internal factors or environmental factors are high. This is modeled by a union operator (OR). We define the internal factors to be the conjunction (AND) of genetic predisposition and genetic test results. In this particular example, the rationale for using a conjunction might be that for some types of cancer, the test might be subject to a high rate of false positives, and so, these results can be offset by low genetic propensity. Environmental factors are aggregated as the disjunction (OR) of amount of smoking, hazardous work risk and the negation, or complement (NOT), of good nutrition. Clearly, this is a gross oversimplification and is included to demonstrate the utility of fuzzy operators in multicriteria decision making more than focusing on reality. In Figure 2.4, we model each of the operators with the corresponding Yager connective (Equation 2.1). The parameters for these four connectives will be labeled w_1 for the top disjunction, w_2 for the conjunction of internal factors, w_3 for the disjunction of external factors and w_4 for the complement.

Given the tree in Figure 2.4, suppose that we have determined the following fuzzy membership values for the leaf nodes: propensity = 0.2, test results = 0.8, smoking risk = 0, job risk = 0.1, and good nutrition = 0.8. That is, a particular patient has a low genetic propensity but a reasonably high likelihood from a test, while living a good lifestyle. If the logical operators are the classical binary ones, then the risk of cancer would be zero for this set, assuming that the fuzzy memberships are turned binary at a 0.5 threshold. The advantage of fuzzy set theory is that the operators that govern complement, disjunction and conjunction can be tailored to reflect different user dispositions. Table 2.1 displays the "cancer risk" output for a few choices of connection parameters. For example, using the parameters on line 3 of Table 2.1, we calculate the *Risk* as:

$$Risk = \min\{1,((1-\min\{1,((1-0.2)^{0.5}+(1-0.8)^{0.5})^2\})^{0.5}+$$
$$(\min\{1,((\min\{1,(0^1+0.1^1)^{1/1}\})^1+((1-0.8^2)^{0.5})^1)^{1/1}\})^{0.5})^2\} = 0.7$$

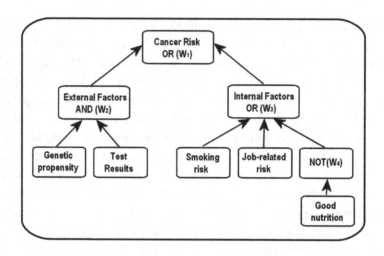

Figure 2.4 Oversimplified tree structure to demonstrate the utility of fuzzy operators

The weights in case 1 of Table 2.1 produce operators that behave like the classical binary ones, and hence produce risk near zero. If the patient or doctor is more aggressive relative to assessing risk, Table 2.1 provides examples that produce low, moderate and even complete risk for those same inputs. While we claim that this flexibility is an advantage of fuzzy set theory, some may argue that it confuses the situation. The message is that no one should use computational or logical operations on data without understanding how these operators combine the data. By studying fuzzy set connectives (as in [Klir and Yuan, 1995]), different degrees of aggressiveness can be quantified and produce meaningful tradeoffs to a patient in this case, or for more general multicriteria decision making processes.

Table 2.1 Cancer risk output for various weights for the input values stated in the text

Parameters	w_1	w_2	w_3	w_4	Risk
1	0.5	0.5	2.0	0.5	0.1
2	1.0	10.0	1.0	0.5	0.3
3	0.5	0.5	1.0	2.0	0.7
4	0.5	0.5	0.5	2.0	1.0

Table 2.2 Risk output for various smoking habits for $w_1=1.0$, $w_2=10.0$, $w_3=1.0$, $w_4=0.5$ and for the input values stated in the text.

Smoking	0.0	0.1	0.2	0.3	0.4	0.5	0.6	0.7	0.8	0.9	1.0
Risk		0.3	0.4	0.5	0.6	0.7	0.8	0.9	1.0	1.0	1.0	1.0

Additionally, for a given choice of parameters, a "what-if" game can be played. In the above example, with the parameters as in case 2 of Table 2.1, we can examine the change in cancer risk given by changing a patient's smoking habits, as displayed in Table 2.2.

2.4.3 Compensatory operators

Of course, the meaningfulness of the results of an analysis as in example 2.1 depends on the faithfulness of the model and the accuracy of assessing the input values. This problem is not specific to fuzzy set theory, but is inherent to all computational paradigms. The discussion here makes no overt claims to accurately model cancer risk, but is only used to demonstrate the flexibility of fuzzy connectives in decision processes.

In Figure 2.4, we might conjecture that the internal and external factors might better be combined in a compensative manner, i.e., more like an average than a union. Besides modeling negation (NOT), disjunction (OR) and conjunction (AND), fuzzy set theory admits mechanisms to model compensatory connections, i.e., aggregation operators where a high value in matching one criterion can compensate to some extent for a low value for another criterion. The simplest of these is called the generalized mean. If $a_1, a_2, ..., a_n$ are the degrees of satisfaction of n criteria, the generalized mean is defined as:

$$h_\alpha(a_1, a_2, \cdots, a_n) = \left(\frac{a_1^\alpha + a_2^\alpha + \cdots + a_n^\alpha}{n} \right)^{1/\alpha} \tag{2.2}$$

where α is a fixed real number. For $\alpha = 1$, this equation implements the arithmetic average, for $\alpha = -1$, we have the harmonic average, and for α converging to 0, Equation 2.2 produces the geometric mean, the nth root of the product of the values. All instantiations of the generalized means produce values between the minimum and maximum of the degrees of satisfaction of the individual criteria. Additionally,

$$\lim_{\alpha \to -\infty} h_\alpha(a_1, \cdots, a_n) = \min\{a_1, \cdots, a_n\} \quad \text{and}$$

$$\lim_{\alpha \to \infty} h_\alpha(a_1, \cdots, a_n) = \max\{a_1, \cdots, a_n\}.$$

This high degree of coverage makes fuzzy set connectives appealing for multicriteria decision making.

In [Krishnapuram and Lee, 1992a, 1992b], Yager unions and intersections, along with generalized means, were used in hierarchical decision networks, and a gradient descent-based training algorithm was created to learn the parameters of the connectives in the network from a set of input/output training data. However, there were fairly cumbersome tests to decide if a node should be a union, intersection or mean (and to flip between them). A more general class of connectives, called fuzzy hybrid operators, combine all three types of linguistic connectives into a single equation. The typical arithmetic and multiplicative hybrid operators are given by:

$$A \oplus_\gamma B = (1 - \gamma)(A \cap B)^{1-\gamma} + \gamma(A \cup B) \tag{2.3}$$

$$A \otimes_\gamma B = (A \cap B)^{1-\gamma} \cdot (A \cup B)^\gamma \tag{2.4}$$

where γ is between 0 and 1 and controls the amount of "mixing" of the union and intersection components, i.e., if γ is close to 0, the hybrids acts like an intersection, near 1 produces a union-like response, and for γ around 0.5, the hybrid takes on the characteristics of a generalized mean.

Zimmermann and Zysno [1980] proposed a hybrid operator for multicriteria aggregation that was modeled after the compensatory nature of human aggregation. This hybrid operator (γ model) is an example of Equation (2.4) and is given by:

$$Y = (\prod_{i=1}^{n} (a_i)^{\delta_i})^{1-\gamma} (1 - \prod_{i=1}^{n} (1 - a_i)^{\delta_i})^{\gamma} \qquad (2.5)$$

where, $a_i \in [0,1]$ are the criteria satisfactions to be aggregated,

$0 \le \gamma \le 1$ is the mixing coefficient, and $\sum_{i=1}^{n} \delta_i = n$. Here, δ_i are weights

associated with each criterion a_i and n is the number of criteria being aggregated.

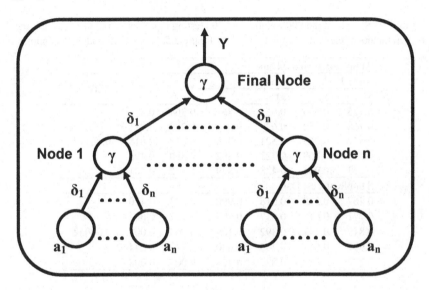

Figure 2.5 A fuzzy aggregation network of multiplicative hybrids used in [Parekh and Keller, 2007].

Krishnapuram and Lee [1992a,b] also developed a back-propagation algorithm to learn the parameters of operators of this type in a network-based decision application. While the algorithm converged, the derivatives were quite messy and as with all such algorithms, convergence could only be guaranteed to a local minimum of a least squares fitness function. Keller *et al.* [1994] extended the approach to additive hybrid networks. In [Parekh and Keller, 2007], particle swarm optimization [Eberhart and Kennedy, 1995; Clerc, 2004] was used to train these aggregation networks. Figure 2.5 displays such a network. The advantage of swarm optimization is that many potential solutions (here, the list of all node parameters) are randomly generated and through individual particle memory and communication between particles, large areas of the optimization search space can be covered while still moving quickly to a (usually very good) local optimum of the fitness function. Additionally, in this case, no derivatives were necessary, since each particle contains all the node parameters and evaluation is performed directly at each time step.

Table 2.3 Sample of the 800 training and 200 testing input/output data for learning the parameters of the nodes in the network of Figure 2.5, where each node has 2 inputs

Sample of Training Data						
Node 1		Node 2		Y	Y'	SSE
a1	a2	a1	a2			
0.425	0.590	0.655	0.861	0.364	0.363	
0.768	0.452	0.629	0.668	0.209	0.211	
0.532	0.053	0.521	0.548	0.018	0.018	0.00175
0.235	0.868	0.722	0.892	0.490	0.492	
0.673	0.925	0.428	0.829	0.542	0.541	
Sample of Test Data						
0.467	0.538	0.518	0.990	0.423	0.420	
0.771	0.678	0.617	0.999	0.621	0.622	
0.810	0.344	0.392	0.109	0.005	0.005	0.00052
0.997	0.644	0.235	0.630	0.242	0.244	
0.272	0.032	0.821	0.528	0.009	0.008	

Example 2.2. As an example, synthetic data was used to verify this approach. Parameters for the multiplicative hybrid operators were randomly generated and assigned to each node. Then, a table of 1000 input values was randomly generated and corresponding outputs were calculated from successive applications of Equation 2.5. The training data consisted of 800 data points and the test data had 200 data points. Table 2.3 shows a sample of the training and test data from one experiment while Table 2.4 shows the original and recovered hybrid parameters. With an easy and effective training mechanism, such fuzzy aggregation networks are attractive tools for hierarchical confidence

fusion. Besides the ability to approximate input/output training data, an additional advantage of these networks is that after training, each node can be associated with a linguistic connective (disjunction, conjunction, mean), based on the corresponding value of γ, and the weights give an indication of the importance of the particular criteria towards the fused result.

Table 2.4 Actual and recovered parameters corresponding to Table 2.3.

	Parameter	Actual	Recovered
Node 1	δ_1	0.440	0.446
	δ_2	1.559	1.553
	γ	0.255	0.341
Node 2	δ_1	0.161	0.163
	δ_2	1.838	1.836
	γ	0.180	0.198
Final Node	δ_1	0.786	0.816
	δ_2	1.213	1.183
	γ	0.0846	0.028

2.5 Fuzzy Relations and Fuzzy Logic Inference

There are times when domain knowledge, and hence, the decision functions, about a particular problem can be best described in terms of linguistic rules. For example, in the cancer risk example (Example 2.1), we might have rules like

> **IF** The Internal Factor Risk is SOMEWHAT LOW and
> The External Factor Risk is LOW
> **THEN**
> The Overall Cancer Risk is LOW.

Concrete applications of using linguistic rules will be shown in Chapters 3 and 5. Traditional crisp expert systems including those that manipulate numeric confidences or probabilities have been around for many years [Ignizio, 1991; Giarratano and Riley, 2005]. Fuzzy logic extends this approach by modeling linguistic propositions, rules and the inference procedure directly with fuzzy sets. In this section we describe the background necessary to understand and construct fuzzy logic inference systems for bioinformatics problems.

Fuzzy logic begins with the concept of a linguistic variable [Zadeh, 1975a,b; Zadeh, 1976]. A linguistic variable is, as its name suggests, a variable whose values are words. For example, the linguistic variable "Age" might take as values "infant", "youth", "adult", "middle-agged", "senior", "elder". With any linguistic variable, there is an underlying domain, X, that will be used to create the meanings for the linguistic values. In our simple illustration above, the underlying domain consists of the real numbers between 0 and 120. Each linguistic value has a fuzzy subset of X that serves as its definition. An example will be given at the end of this section.

Once we have this fundamental concept, we can build the machinery necessary for fuzzy logic inference. In what follows, let $X, X_1, X_2, ..., X_n,$ and Y be domains, $U, U_1, U_2, ..., U_m,$ and V be linguistic variables, $A, A_1, ..., A_n$, $B, B_1, ..., B_k$ be the fuzzy sets that model linguistic values over respective domains. An atomic proposition in fuzzy logic is a statement of the form "U is A", where U is the name of a linguistic variable and A is the name of a linguistic value, i.e., it is the name of a fuzzy subset of the domain X. In the above rule, one atomic proposition is "Internal Risk Factor is Somewhat Low".

The conjunctive proposition between two fuzzy set can be written as follows [Klir and Yuan 1995]:

$$U_1 \text{ is } A_1 \text{ and } U_2 \text{ is } A_2$$

where the U_i are linguistic variables over domains X_i; and where $A_i(x_i)$ are linguistic values represented by fuzzy sets on those domains.

The result of this operation is a fuzzy relation of the cross product domain, based on U_1 and U_2 which is called the Cylindrical Closure of the fuzzy sets A_1 and A_2. A fuzzy relation so referenced is just a fuzzy subset of $X_1 \times X_2$. The Cylindrical Closure can be viewed as the intersection of the extension of each fuzzy set to the cross product domain $X_1 \times X_2$, i.e., a fuzzy subset, $A_1 \times A_2$, of $X_1 \times X_2$ where $A_1 \times A_2(x_1, x_2) = A_1(x_1) \wedge A_2(x_2)$. Here we use minimum as the intersection operator, but note that any fuzzy intersection operator could be used.

A condition proposition, or fuzzy implication, between two fuzzy propositions is written as

IF U is A **THEN** V is B

where U and V are linguistic variables that have elements $x \in X$, and $y \in Y$ respectively; and where $A(x)$, $B(y)$ are linguistic values represented by fuzzy sets on those elements. The definition of an implication proposition is a fuzzy relation R between X and Y, based on U and V, that can take many forms in combining the input fuzzy sets. Three common definitions used in many fuzzy rule systems are:

The Lukasiewicz implication (Zadeh's original implication operator):

$$R_z(x, y) = \min(1, 1 - A(x) + B(y))$$

Correlation min implication:

$$R_{cm}(x, y) = \min(A(x), B(y))$$

Correlation product implication:

$$R_{cp}(x, y) = A(x) * B(y)$$

Note that a fuzzy implication proposition is just a (fuzzy) rule. The Compositional Rule of Inference or Generalized Modus Ponens can now be described to combine a fuzzy rule and a linguistic proposition.

The Compositional Rule of Inference is [Zadeh, 1973]:

Rule: **IF** U is A **THEN** V is B

Fact: U is A'

Conclusion: V is B'

where the expression of conclusion is the composition operation:

$$B'(y) = A'(x) \circ R(x, y)$$

Here, $R(x, y)$ is the chosen translation of the fuzzy implication. The composition operation is defined as:

$$B'(y) = A'(x) \circ R(x, y) = \sup_{x \in X} \min\{A'(x), R(x, y)\}$$

where "sup" is the supremum of the set, i.e., the least element that is greater than or equal to each element in the set (the Max if all sets are finite).

Rarely will rules only have one antecedent clause. Rules with multiple antecedent clauses pose no conceptual problem. The Compositional Rule of Inference with multiple antecedent clauses is the following:

Rule: **IF** U_1 is A_1 and U_2 is A$_2$ and ...and U_n is A$_n$ **THEN** V is B.

Fact: U_1 is A'_1 and U_2 is A'_2 and ... and U_n is A$'_n$.

Conclusion: V is B'.

The first step in this case is to find the cylindrical closure, $A_1 \times A_2 \times \cdots \times A_n$, of the n antecedent clauses, i.e., the intersection of the extensions of all these fuzzy sets to the domain $X_1 \times X_2 \times \cdots \times X_n$. Once computed, the chosen definition for implication can be applied to the rule:

IF $<U_1, U_2, \cdots, U_n>$ is $A_1 \times A_2 \times \cdots \times A_n$ **THEN** V is B.

This produces a fuzzy relation R between $X_1 \times X_2 \times \cdots \times X_n$ and Y, i.e., a fuzzy subset of $X_1 \times X_2 \times \cdots \times X_n \times Y$. Finally, the fuzzy conclusion can be drawn with the compositional rule of inference as

$$B'(y) = A'_1 \times \cdots \times A'_n(x_1, \ldots, x_n) \circ R(x_1, \ldots, x_n, y)$$

The Compositional Rule of Inference with several rules takes the following form:

Rule 1: **IF** U_1 is A_{11} and ...and U_n is A_{1n} **THEN** V is B_1

Rule 2: **IF** U_1 is A_{21} and ...and U_n is A_{2n} **THEN** V is B_2

$\vdots$

$\vdots$

Rule k: **IF** U_1 is A_{k1} and ...and U_n is A_{kn} **THEN** V is B_k

Fact: U_1 is A'_1 and U_2 is A'_2 and ... and U_n is A'_n

Conclusion: V is B'

Each rule is translated as above to form $R_i(x_1, \ldots, x_n, y)$ and then the compositional rule of inference is applied to that rule with the fact proposition to obtain $B_i'(y) = A'_{i1} \times \cdots \times A'_{in}(x_1, \ldots, x_n) \circ R_i(x_1, \ldots, x_n, y)$. These partial conclusion expressions are aggregated into a single output fuzzy set by either

$$B'(y) = \sum_{i=1}^{k} B_i'(y) \text{ or}$$

$$B'(y) = \max_{i=1}^{k}\{B_i'(y)\}.$$

Note that the first expression may exceed 1 for particular values of y and hence, not formally be a fuzzy subset of Y. However, it is easy to normalize. In fact, this formula is popular in those cases like fuzzy control where the output fuzzy set needs to be converted to a single numeric value. This process is known as defuzzification. The most common form is centroid defuzzification:

$$\bar{y} = \frac{\sum_{y \in Y} y \cdot B(y)}{\sum_{y \in Y} B(y)}.$$

While the above development handles the general case of fuzzy inference, in most applications of fuzzy rule-based systems, the inputs are not actually fuzzy sets themselves, but crisp values in their respective domains. For example, in a fuzzy system to perform classification, the inputs may be values of features extracted from the objects to be classified. The rules may contain propositions like "Feature 1 is LOW", "Class 1 confidence is HIGH", etc., indicating the uncertainty in the decision process. However, in application, given an object to classify, Feature 1 is normally a real number, x_1. The standard method to convert it to a fuzzy set for the inference process is to create a (crisp) set that is 1 for x_1 and zero everywhere else in the domain X_1. This makes the firing of the rules particularly simple [Klir and Yuan 1995]. Alternately, a simple triangular set, a pi-function, or any membership function can be centered at x_1 to explicitly model the uncertainty in the feature extraction (see Figure 2.2). This process of converting measured crisp inputs into fuzzy sets for inference purposes is known as fuzzification. It may seem artificial at times, but the rule clauses themselves describe the uncertainty and variability in the problem domain, so, the exact form of fuzzification is less critical.

The system of inference described above is referred to as a Mamdani-Assilion or MA fuzzy rule system [Mamdani, 1977]. An alternate formulation, denoted as a Takagi-Sugeno-Kang (TKS) system [Takagi and Sugeno, 1985; Sugeno and Kang, 1988] only modifies the membership functions in the consequent clause. It was developed for control applications where the output of the rule firing should be a function of the set of crisp input values. Instead of a general fuzzy set B of Y, the output of each rule is a specific function of the real inputs. The antecedent part of each rule, R_i, is matched as in the MA approach, but the output then becomes $y_i = A'_{i1} \times \cdots \times A'_{in}(x_1,\ldots,x_n) \cdot f_i(x_1,\ldots,x_n)$. The weighted average of this set of k values is used as the system output. One of the main motivations to recast fuzzy inference in this way is that stability theory for fuzzy controllers could be developed [Passino, 1998; Verbruggen and Babuska, 1999]. For the purposes of this book, either method can be used to produce similar results for bioinformatics problems. The choice is really in the description of the consequent clause, as will be demonstrated in the examples below.

Fuzzy logic systems are quite powerful and have been used to in many applications from non-linear control to classification. However, a common question is often asked: Where do the rules come from? Much like our discussion of membership functions, sometimes the rules come from experts. The person who has controlled a complex piece of equipment can linguistically describe his or her reactions to a variety of input conditions. Think about a simple task of balancing a broom in the palm of your hand. While you may not be able to solve the equations for motion in your head, at any instant, you can see roughly the angle the broom makes with the desired vertical orientation, say in terms of words like "big", "medium", "small", etc., and you can "feel" the rate at which the broom is moving, either up or down and quantized in a similar way to the angle. (Note that here we only consider the broom falling away from us, through both directions should be taken into account.) It wouldn't take long for you to come up with rules like "IF the broom's angle is MEDIUM and the broom is falling away from me SLOWLY THEN push my hand forward FAIRLY FAST". Balancing the "inverted pendulum" on a motorized cart was one of the early demonstrations of fuzzy logic rule-based control.

In cases where training data is available, fuzzy rules can be learned, many times through the use of clustering algorithms or other computational intelligence techniques like neural networks, evolutionary computation, swarm intelligence, etc. (for example, see [Pedrycz and Gomide, 1998, Zurada *et al.*, 1994, Fogel and Robinson, 2003]). Learning the rules (and their membership functions) is treated as an optimization problem; the performance of the rule system on the training data is the function that needs to be maximized. There are many tools available to manually build or to learn fuzzy rule systems. In fact, in the example given below, Matlab contains a fuzzy logic toolbox containing MA, TKS and a "neuro-fuzzy" implementation of the TSK model (called ANFIS [MathWorks, 1995]) that supports learning both the antecedent membership functions and the consequent function parameters.

Fuzzy rule-based systems clearly generalize standard expert systems. In the inverted pendulum example, an equivalent crisp rule control system could be developed by quantizing the angle and the rate of motion into small intervals and building a rule for all pairs of intervals, the output of each rule being a set velocity of cart motion. How fine does the quantization need to be? That depends on how smooth you desire the control to be. Fuzzy rule systems, by their very construction, are great interpolators and so, few rules are usually needed when compared to crisp expert systems. Should fuzzy rules always be used? If the data is purely symbolic in nature, then fuzzy logic certainly doesn't apply. Probabilities can be associated with crisp rules and uncertainty can be updated along with rule firing. Bayesian networks, or more generally, belief networks offer alternative ways to encode and manage probabilistic uncertainty in hierarchical frameworks. The choice of model should always be dictated by the form of the problem, the nature of the uncertainty, the ease of use of the particular formulation, and the meaningfulness of the results.

In Chapters 3 and 5, we will show specific applications of fuzzy rules to bioinformatics. Here, we close with a straight forward example of a fuzzy inference system from [Wang *et al.*, 2006].

Example 2.3. This application involves softening the output of the Short Physical Performance Battery (SPPB) test, a series of timed physical activities that have been created to evaluate, discriminate, and predict physical functional performance for both research and clinical purposes, primarily for physically impaired older adults. The original scoring system of the SPPB test uses crisp time boundaries to assign the subject to discrete classes of performance. The crisp (and somewhat arbitrary) nature of the thresholds can easily produce anomalies. The SPPB test measures balance, gait, strength, and endurance. Although it is a timed performance test, each subtask score is an integer value in the range 0-4. A score of 0 indicates the inability to complete the task in a nominal time frame while categories 1-4 are assigned to the corresponding quartiles of time needed to perform the action. The original scoring for the SPPB standing test is shown in Table 2.5 [Guralnik *et al.*, 1994].

Table 2.5 Scoring performance on tests of standing balance

Score	Side by side stand	Semi-tandem stand	Full tandem stand
0	t< 10 sec.	Not attempted	Not attempted
1	t=10 sec.	t< 10 sec.	Not attempted
2	t=10 sec.	t=10 sec.	t< 3 sec.
3	t=10 sec.	t=10 sec.	3sec.<=t<10sec.
4	t=10 sec.	t=10 sec.	t=10 sec.

In [Wang *et al.*, 2006], rules and the membership functions for the linguistic values were constructed manually with input from nurses. The set of fuzzy rules for Standing Test performance is:

      1. **IF** (Side-by-Side_Stand_Time is SHORT) **THEN** (Standing_Test_Performance is VERY_POOR)

      2. **IF** (Side-by-Side_Stand_Time is LONG) and (Semi-Tandem_Stand_Time   is   SHORT)   **THEN** (Standing_Test_Performance is POOR)

      3. **IF** (Side-by-Side_Stand_Time is LONG) and (Semi-Tandem_Stand_Time is LONG) and (Full-Tandem_Stand_Time is SHORT) **THEN** (Standing_Test_Performance is OK)

4. **IF** (Side-by-Side_Stand_Time is LONG) and (Semi-Tandem_Stand_Time is LONG) and (Full-Tandem_Stand_Time is MEDIUM) **THEN** (Standing_Test_Performance is GOOD)

5. **IF** (Side-by-Side_Stand_Time is LONG) and (Semi-Tandem_Stand_Time is LONG) and (Full-Tandem_Stand_Time is LONG) **THEN** (Standing_Test_Performance is EXCELLENT).

Membership functions were modeled by either triangles and trapezoids or smooth curves, in this case, chosen heuristically to reflect common sense. As an example, the membership functions for Short, Medium, and Long for the linguistic variable Full-Tandem Stand are shown in Figure 2.6.

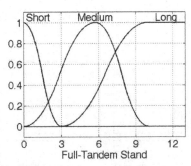

Figure 2.6 Membership functions for Full-Tandem Stand, used in the fuzzified scoring rule-based SPPB system.

The system was implemented in Matlab [MathWorks, 1995] using both an MA fuzzy set output and a functional TSK output format (the output function for each rule is just the class label value, 0-4). Figure 2.7 displays one implementation of an MA system response when the side by side stand is 10 seconds, semi-tandem stand is 10 seconds, and full tandem stand is 9 seconds. Here, the defuzzified output is 3.2, close to the crisp output of 3 in this case. The fuzzy system provides a smoother transition from one category to the next as the times change. One goal of that project is to do frequent passive monitoring of elders to detect gradual changes in their physical performance, and such a fuzzy system provides a higher degree of specificity to do early detection of functional decline.

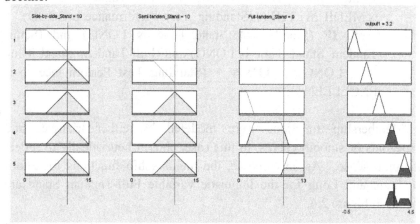

Figure 2.7 Matlab implementation of simple MA rule system for PSSB scoring

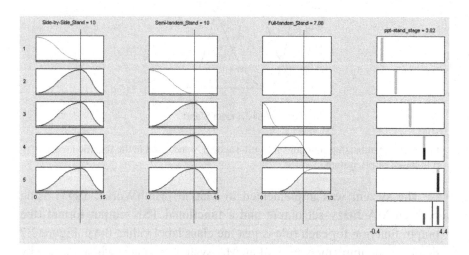

Figure 2.8 Matlab implementation of simple TSK rule system for fuzzy PSSB scoring.

Figure 2.8 shows a similar configuration for a TSK version of the rule base, with smooth membership functions. Particularly with small rule bases, the performance is not overly sensitive to the form of the precise definition of the membership functions.

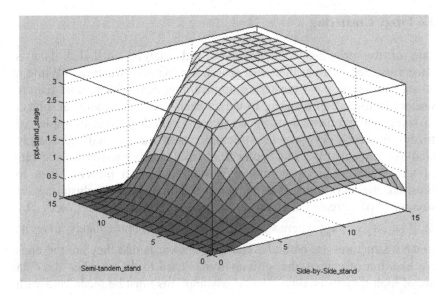

Figure 2.9 Output surface of fuzzy SPPB rules system with respect to Semi Tandem Stand time and Side by Side Stand time.

Finally, Figure 2.9 shows the complete output surfaces obtained by varying two of the three input values across their entire respective ranges. The figure clearly shows the smoothness of the output function to small changes.

2.6 Fuzzy Clustering

One of the principal tools to mine and analyze unlabeled data is clustering, that is, algorithms that search for "natural structure". While it is not necessary, most clustering applications deal with sets of feature vectors in Euclidean d-space, in which each vector represents an object in some real problem domain. Chapters 4 and 5 present examples of fuzzy clustering in protein structure classification and microarray analysis, respectively. In Chapter 3, objects to be clustered will be sets of terms coming from an ontology that describe a protein. But to illustrate the algorithms, it is convenient to first describe clustering for vectors of real numbers. We denote the set by $X = \{\mathbf{x}_1, \mathbf{x}_2, \ldots, \mathbf{x}_n\}$ where $\mathbf{x}_k \in R^d$. The question immediately arises as to what structures are natural. The easy answer is that they are the ones we like, but of course, that glib answer is hard to define when the data are of high dimensionality. Hence, assumptions must be made, and with each assumption, we place constraints on the groupings allowed by automated techniques. For example, it is completely reasonable to assert that points in feature space that are "close" to each other end up in the same cluster. Close is normally defined by a distance metric in R^d. Certainly the choice of distance measure strongly influences the resultant grouping of data. The standard Euclidean distance $d^2(\mathbf{x}, \mathbf{y}) = (\mathbf{x} - \mathbf{y})^t(\mathbf{x} - \mathbf{y})$, the dot product of the difference between the two vectors, favors groups of vectors that are hyperspherical. Different choices of distance functions or, as we will see, dissimilarity measures give rise to alternate definitions of closeness of objects for clustering approaches.

All clustering is based on the concept of a C-partition of the data set *X*. A partition of n data points into *C* clusters is defined by a partition matrix $U = \{u_{ik}\}$, where $0 \leq u_{ik} \leq 1$ is the degree of data point $\mathbf{x}_k$ belonging to cluster A_i, subject to the constraint that the total degree of a data point belonging to all clusters being one, that is,

$$\sum_{i=1}^{C} u_{ik} = 1 \quad \text{for all } k. \tag{2.6}$$

For simplicity, we will also call u_{ik} the degree of membership of data point $\mathbf{x}_k$ in cluster A_i. Note that in the crisp case, each $\mathbf{x}_k$ will be assigned to one and only one cluster A_i. In other words, for each $k = 1$, ..., n, $u_{ik} = 1$ for some i between 1 and C and $u_{jk} = 0$ for all other cluster indices j.

2.6.1 Fuzzy C-Means

The Fuzzy C-Means (FCM) [Bezdek et al., 1999] is a scheme to partition a set of data into a predefined number of clusters taking into account the uncertainty of cluster assignment. It effectively allows for sharing of objects between clusters. In this approach, each cluster is represented by an exemplar (or prototype or cluster center). Let $\mathbf{v}_i$ be the prototype of cluster A_i and let V be the set of all C cluster centers. The objective of FCM is to minimize the following criterion function:

$$J(U,V) = \sum_{k=1}^{n} \sum_{i=1}^{C} (u_{ik})^m d^2(\mathbf{x}_k, \mathbf{v}_i), \qquad (2.7)$$

subject to the constraint that $\sum_{i=1}^{C} u_{ik} = 1$ for all k. The constraint is necessary to guard against the trivial solution, i.e., setting all cluster memberships to zero. Here, the parameter m is called the fuzzifier. This is because larger values of m favor more "fuzzy" partitions, that is, more similar degrees of membership of a data point in all clusters. Performing this minimization leads to the following two equations expressing necessary conditions for a minimum.

For point prototypes, that is, each cluster is represented by a single vector in the feature space, these prototypes must have the form

$$\mathbf{v}_i = \frac{\sum_{k=1}^{n} (u_{ik})^m \mathbf{x}_k}{\sum_{k=1}^{n} (u_{ik})^m} \qquad (2.8)$$

The necessary condition on the memberships values at a minimum of the criterion function are

$$
u_{ik} = \frac{\left(\dfrac{1}{d(\mathbf{x}_k, \mathbf{v}_i)}\right)^{\frac{2}{m-1}}}{\displaystyle\sum_{j=1}^{C}\left(\dfrac{1}{d(\mathbf{x}_k, \mathbf{v}_j)}\right)^{\frac{2}{m-1}}}
\qquad (2.9)
$$

Note that Equations 2.8 and 2.9 are coupled in the sense that the partition memberships are needed to compute the prototypes and the prototypes are required to update the memberships. The FCM algorithm performs an iterative technique called Alternating Optimization (AO) where cluster memberships and cluster centers are alternately updated in each iteration. There is a technical detail in the event that one of the data points coincides with a cluster center. In that case, we assign complete membership of that point to the corresponding cluster and zero membership to all other clusters. The algorithm can be summarized as follows.

Let $X = \{\mathbf{x}_1, \mathbf{x}_2, ..., \mathbf{x}_n\}$ where $\mathbf{x}_k \in R^d$ be the set of vectors to be clustered.

    Initialization: Set
        C, the number of clusters desired
        m, the fuzzifier
        ε, the convergence threshold
        $V^{(0)} = \left\{\mathbf{v}_1^{(0)}, ..., \mathbf{v}_C^{(0)}\right\}$ an initial set of cluster centers

           //Note: The $\mathbf{v}_i^{(0)}$ can be chosen randomly from
               X or through other mechanisms //
        Set $t = 0$
    REPEAT
        DO FOR each $k = 1, ..., n$
        IF $d(\mathbf{x}_k, \mathbf{v}_i) = 0$ for some i
        THEN
           set $u_{ik}^{(t)} = 1$ and $u_{jk}^{(t)} = 0$ for $j \neq i$
        ELSE

> Estimate $u_{ik}^{(t)}$ from Equation 2.9.
> > ENDIF
> END FOR
> Set $t \leftarrow t+1$
> > Using $U^{(t-1)}$, estimate $V^{(t)}$ from Equation 2.8.

$$\text{UNTIL } \sum_{i=1}^{C} \left\| \mathbf{v}_i^{(t)} - \mathbf{v}_i^{(t-1)} \right\| < \varepsilon$$

where $\| * \|$ is any vector norm (like Euclidean).
//Note: there are other stopping criteria, including number of iterations, but this is the most common//

If the cluster memberships are required to be binary, i.e., the clustering algorithm is to build a crisp C-partition of X, then the above AO algorithm reduces to the Crisp or Hard C-Means (HCM) with the two steps in the UNTIL loop: 1. assign each vector to the cluster with the closest cluster center, and 2. compute new cluster centers as the means of the vectors assign to the respective clusters [Theodoridis and Koutroumbas, 2006].

Table 2.6 Fifteen point butterfly data set

Point	x_i	y_i
1	0	0
2	0	2
3	0	4
4	1	1
5	1	2
6	1	3
7	2	2
8	3	2
9	4	2
10	5	1
11	5	2
12	5	3
13	6	0
14	6	2
15	6	4

Example 2.4. As a simple example, consider the 15 point 2-dimensional data set, called the butterfly data, listed in Table 2.6 and shown graphically in Figure 2.10.

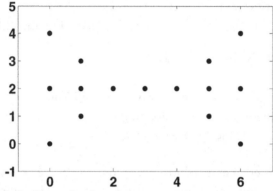

Figure 2.10 Butterfly data set for clustering.

Visually, there looks like 2 clusters with a bridge point (#8) between the symmetric clusters. Running the Hard 2-Means, using randomly chosen initial cluster centers, the final cluster memberships are displayed graphically in Figure 2.11, whereas the final cluster centers coordinates, $\{ (1, 2)^t, (5.3, 2)^t \}$, are shown in Figure 2.12. The bridge point had to be assigned to one of the clusters, in this case, the left cluster. The crisp assignment required by the HCM clearly affects the locations of the cluster centers.

Next, we ran the Fuzzy 2-Means on this data with $m = 2$. Figures 2.13 and 2.14 show the fuzzy memberships and locations of the final cluster centers for the $m = 2$ case. The bridge point now is shared between the two clusters with final membership of 0.5 in each.

To show one of the properties of the FCM, namely that it converges to the HCM results as m approaches 1, we ran the Fuzzy 2-Means on the butterfly set with $m = 1.25$. Figure 2.15 corresponds to fuzzy memberships for this case. Memberships except for the bridge point are quite crisp. That vector is still shared between the two clusters with final memberships almost 0.5 in each (slightly higher in the second cluster for this run). The locations of the final cluster centers are in essentially the same place as for $m = 2$.

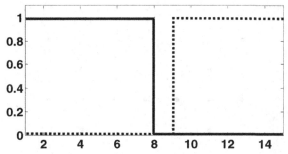

Figure 2.11 Hard 2-Means memberships of butterfly points by point index. Note that the bridge point (#8) is assigned to cluster 1, represented by the solid line.

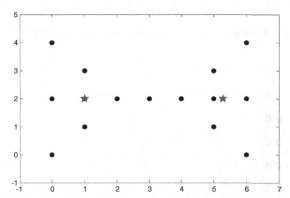

Figure 2.12 Butterfly data and Hard 2-Means cluster centers (shown as stars).

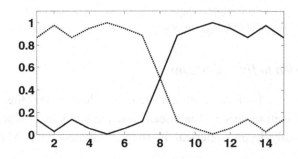

Figure 2.13 Fuzzy 2-Means final memberships for $m = 2$ of butterfly points by point index. Note that the bridge point (#8) is shared equally by the two clusters, that is $u_{18}=u_{28}=0.5$.

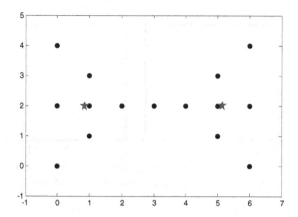

Figure 2.14 Butterfly data and Fuzzy 2-Means cluster centers { $(0.85, 2.0)^t$, $(5.15, 2.0)^t$ } for $m = 2$.

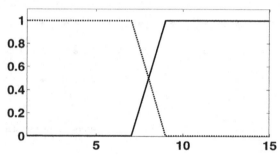

Figure 2.15 Fuzzy 2-Means memberships of butterfly points by point index for $m=1.25$. Note that the bridge point (#8) is again shared almost equally by the two clusters.

2.6.2 Extension to fuzzy C-Means

The choice of distance function actually determines the retrieved geometry of feature space. Euclidean distance tends to produce clusters that are roughly hyper-spherical in shape. Consider the Mahalanobis distance:

$$d^2(\mathbf{x}_k, \mathbf{v}_i) = (\mathbf{x}_k - \mathbf{v}_i)^t \Sigma_i^{-1} (\mathbf{x}_k - \mathbf{v}_i) \qquad (2.10)$$

Here, Σ_i represents the estimated "fuzzy" covariance matrix for the ith cluster [Bezdek *et al.*, 1999; Theodoridis and Koutroumbas, 2006].

Using this distance metric, which changes for each cluster, the resulting clusters can assume hyper-ellipsoidal shapes. In the AO algorithm of the FCM, these matrices need to be estimated after the cluster centers are computed. The covariance approximation equations are

$$\Sigma_i = \frac{\sum_{k=1}^{n}(u_{ik})^m (\mathbf{x}_k - \mathbf{v}_i)(\mathbf{x}_k - \mathbf{v}_i)^t}{\sum_{k=1}^{n}(u_{ik})^m} \tag{2.11}$$

Scaling the cluster-specific Mahalanobis distance by the dth-root of the determinant of Σ_i makes the resultant FCM algorithm, called the GK-FCM, more conducive to find hyper-ellipsoidal shaped clusters of different sizes [Gustafson and Kessel, 1979; Bezdek *et al.*, 1999].

Example 2.5. To show the advantage of picking the right distance measure, we constructed a data set that consists of three Gaussian clouds with different parameters and mixture probabilities (given by number of points in the "cluster"), and displayed in Table 2.7.

Table 2.7 Parameters for Three Gaussian Clouds data set.

Mean Vector	Covariance Matrix	Number of points
$\mu_1 = \begin{pmatrix} 5 \\ 5 \end{pmatrix}$	$\Sigma_1 = \begin{pmatrix} 5 & 0 \\ 0 & 5 \end{pmatrix}$	100 (o)
$\mu_2 = \begin{pmatrix} -10 \\ 3 \end{pmatrix}$	$\Sigma_2 = \begin{pmatrix} 5 & 4 \\ 4 & 5 \end{pmatrix}$	250 (+)
$\mu_3 = \begin{pmatrix} -3 \\ 6 \end{pmatrix}$	$\Sigma_3 = \begin{pmatrix} 18 & -16 \\ -16 & 18 \end{pmatrix}$	150 (×)

Figure 2.16 shows what it looks like if it were labeled data, i.e., if we have knowledge of the construction of the data set, whereas Figure 2.17 displays the data set as a clustering algorithm would "see" it.

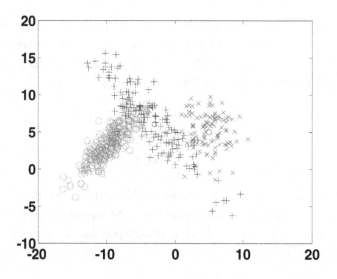

Figure 2.16 Three Gaussian clouds displayed as separate classes.

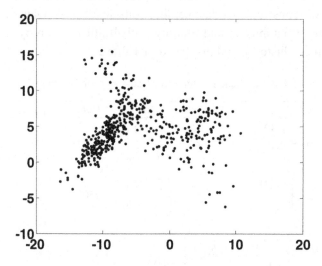

Figure 2.17 Three Gaussian clouds for a clustering exercise.

Now, Figures 2.18, 2.19, and 2.20 show the final crisp partitions for the HCM, the FCM and the GK-FCM. The HCM and the FCM use Euclidean distance. The crisp labels are assigned by picking the cluster with maximal membership for each point. Here, both the HCM and

FCM fail to find the elliptical clusters because Euclidean distance favors circular clusters. The GK-FCM pretty accurately recovers the "correct" structure.

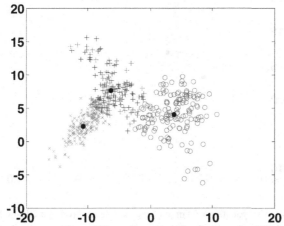

Figure 2.18 Final crisp partition of Three Gaussian Clouds from the HCM with Euclidean distance.

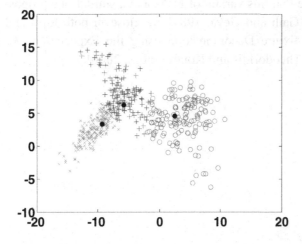

Figure 2.19 Final crisp partition of Three Gaussian Clouds from the FCM with Euclidean distance.

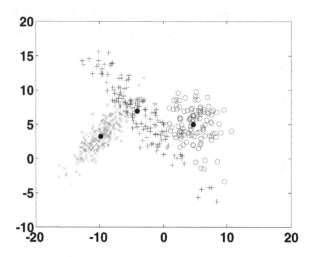

Figure 2.20 Final crisp partition of Three Gaussian Clouds from the GK-FCM with scaled Mahalanobis distance.

We note that this variant of FCM and a similar one proposed by Gath and Geva [Gath and Geva, 1989] are close in both form and results to Gaussian Mixture Decomposition using the Expectation Maximization algorithm [Theodoridis and Koutroumbas, 2006].

2.6.3 Possibilistic C-Means (PCM)

The FCM clustering algorithm ameliorates the problem of crisp assignment of vectors to particular clusters when the features possess ambiguity.   Vector memberships are shared among clusters (the memberships of a given vector must sum to one). When vectors are of high dimension, this fuzzy partition is not only useful in finding strong elements in a cluster (close to binary memberships) but also for detecting objects that lie in an overlapped region when memberships in multiple clusters are close to equal. This was used in [Pal *et al.*, 2005] to find proteins that had been incorrectly annotated in the Gene Ontology (see Chapter 3). However, there is no reason that memberships of a given feature vector should always sum to one. This is a definition within crisp clustering and is required in the FCM to avoid the trivial solution (all memberships equal zero) in minimizing the criterion function.   As mentioned earlier, there are proteins that belong to multiple groups. Under crisp algorithms, such proteins must be placed entirely in one cluster. The FCM can spread the amount of belonging across clusters somewhat better, but still does not capture this condition. Also, since features are measured values, it is often the case that errors in feature extraction occur, resulting in outliers, i.e., points that really do not belong to any cluster. Crisp approaches have no choice but to dump them into a single group and fuzzy algorithms can at best force their memberships close to $1/C$. If the number of points being clustered is not huge, this can result in making noticeable changes in the cluster prototypes.

Krishnapuram and Keller [Krishnapuram and Keller, 1993; Krishnapuram and Keller, 1996] found a way to relax the sum constraint while avoiding the trivial solution. It was done by changing the criterion function, resulting in a clustering technique called the Possibilistic C-Means (PCM). Their criterion function is

$$J(U,V) = \sum_{k=1}^{n}\sum_{i=1}^{C} u_{ik}{}^{m} d^{2}(\mathbf{x}_{k},\mathbf{v}_{i}) + \sum_{i=1}^{C}\eta_{i}\sum_{k=1}^{n}(1-u_{ik})^{m} \qquad (2.12)$$

where the η_i are appropriately chosen or estimated values [Krishnapuram and Keller, 1993; Krishnapuram and Keller, 1996].

The first term is the same as in FCM whereas the second term has the effect of trying to keep cluster memberships high. The necessary conditions to minimize Equation 2.12 now become

$$u_{ik} = \frac{1}{1 + \left(\dfrac{d(\mathbf{v}_i, \mathbf{x}_k)^2}{\eta_i} \right)^m} \tag{2.13}$$

with the condition on the cluster centers identical to Equation 2.8.

Example 2.6. Figure 2.21 shows the final PCM cluster centers $\{(1.04, 2)', (4.96, 2)'\}$ for the butterfly data. These fuzzy clusters are completely symmetric, possible because the constraint that the memberships for each point need to sum to 1 across the clusters is eliminated in the PCM. In this case, the bridge point received low and equal memberships of 0.12 in both clusters and the cluster centers move to symmetric positions.

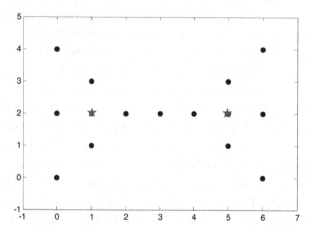

Figure 2.21 The butterfly data and final PCM cluster centers. Symmetric membership functions are produced by the PCM.

The PCM is very robust to outliers [Krishnapuram and Keller, 1996]. To demonstrate this capability, a 16th point was added to the butterfly data set. This point $(3,10)'$ can be considered an outlier since it is far from either of the two more recognizable clusters. This outlier causes serious problems for the Hard 2-Means as seen in Figure 2.22. The final crisp partition splits the data horizontally.

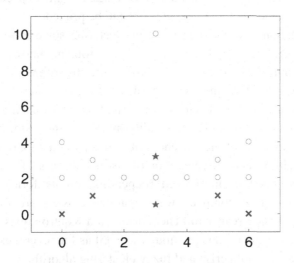

Figure 2.22 Final crisp partition and cluster centers of the HCM on the butterfly+outlier data

The FCM does better, generating a 0.5 membership for both the bridge point and the outlier. The final cluster centers $\{(0.95, 2.35)^t,$ $(5.05, 2.35)^t\}$ are shifted vertically (refer to Figure 2.14) due to the 0.5 outlier membership in both clusters. Using the final FCM cluster centers as initialization for the PCM, as suggested in [Krishnapuram and Keller, 1993], the Possibilistic 2-Means generates the same final cluster centers as in Figure 2.21 because it produces a membership of 0.01 in each cluster for the outlier point.

The solution to the minimization of the PCM effectively decouples clusters, allowing for high or perfect belonging of a vector to multiple groups or for very low membership in all clusters. For this reason, the u_{ik} values are referred to as typicalities instead of memberships, i.e., they represent how similar or typical a vector is of the prototype (cluster center). The PCM tends to search for dense regions of feature space and can have the property that more than one cluster center, and hence the clusters themselves, end up identical. This has been cited as a bad property by some [Barni *et al.*, 1996], but Krishnapuram and Keller argue that it is a good trait in situations where the exact number of clusters is unknown. A larger number of clusters than "expected" can be specified as C and then identical clusters can be pruned.

This situation is a classical issue in clustering since most clustering approaches require knowledge of C. The problem, known as cluster validity, is normally attacked by running the algorithm on a data set several times, varying the number of clusters. A number (called a validity measure) is calculated from the final output of each run. While there are numerous examples of validity measures, most (for object data clouds) are functions of the closeness of points assigned to, or shared by, a particular cluster and the separation of distinct clusters. The measure is usually maximized or minimized (depending on its form) when the "correct" number of compact, well-separated clusters are found. We direct the interested reader to [Theodoridis and Koutroumbas, 2006] for the general concepts of cluster validity as well as for several examples of such measures for both crisp and fuzzy clustering algorithms.

While the algorithms considered above require object data (vectors in R^d), some data sets have the property that only relational information is known. For example, in Chapter 3, we will discuss an application involving the clustering of proteins described by their Gene Ontology annotations. A "distance", or more generally a dissimilarity, will be calculated between these sets of annotations, but these numbers are not derived from vectors of real features. Hence, the data, and algorithms that handle it, are called relational. Most clustering algorithms have relational duals. One such variation of FCM, called Non-Euclidean Relational Fuzzy C-Means (NERFCM) will be developed in Chapter 5 and applied to the problem stated above.

2.7 Fuzzy K-Nearest Neighbors

In the previous section, we discussed various methods to look for structure in sets of unlabeled vectors. In many applications of bioinformatics, like protein structure prediction, we are trying to assign known labels to test data. In these cases, we assume that we have training sets of patterns that represent the various classes (labels) under consideration. The task now is to find a mapping, called a classifier, from the set of new samples into the set of class labels. As with clustering, there are crisp, probabilistic, fuzzy and possibilistic classification models. This section defines a simple, yet powerful family of algorithms to dynamically build a classification mapping that will be used later in chapter 4. This family is referred to as k-nearest neighbor (k-NN) algorithms.

Suppose you wanted to know what the weather was going to be tomorrow and I said "same as today". This seemingly over-simplistic response would be surprisingly good over the long haul. If I wanted a little more evidence, I might use today's weather and that of the previous two days, looking for a two-out-of-three consensus. If I found none, I would pick today's condition over the more remote conditions. Our simple weather example underscores the concepts in both the crisp and fuzzy k-NN algorithms.

As before, suppose $X = \{\mathbf{x}_1, \mathbf{x}_2, ..., \mathbf{x}_n\}$ where $\mathbf{x}_j \in R^d$ is the set of feature vectors, except now we assume that each $\mathbf{x}_j$ has class labels, u_{ij}, $i = 1, ..., C$. The set X is called the labeled training data for the classifier. In the crisp case, $u_{ij} = 1$ for only one class i, and is zero elsewhere. This signifies that $\mathbf{x}_j$ represents an object sampled from class i. By relaxing the binary constraint to have $\sum_{i=1}^{C} u_{ij} = 1$ for each j, or even just requiring $0 \leq u_{ij} \leq 1$, we produce fuzzy or possibilistic labels, respectively, on the training data.

The classical crisp k-NN goes like this. Given an input vector $\mathbf{x}$ to classify, find the closest k neighbors in the training data (pick your favorite distance or dissimilarity measure). Assign $\mathbf{x}$ to the class with the majority label among the k nearest neighbors. Ties are broken arbitrarily. The simplest case is when $k = 1$, called the nearest neighbor or the 1-NN algorithm – like our first weather predictor. Perhaps surprisingly, Cover and Hart [1967] proved that in the limit, the error rate of the 1-NN converged to a value that is not more than twice the optimal Bayes error rate. Of course, we don't live in the limit case, and so, very little theory can be applied to predict the finite performance of the k-NN. However, its simplicity and decent (sometimes, very good) results has made it a very common vector classifier.

In the "more sophisticated" weather predictor, a three-way tie was broken by picking the label (weather conditions) of the closest point. There have been numerous extensions to the k-NN algorithm that make use of the individual distances, $d_j(\mathbf{x}) = d(\mathbf{x}, \mathbf{x}_j)$ for $j=1,\ldots,k$ [Dasarathy, 1991]. Here, we show how both the distances and fuzzy (or possibilistic) labels are combined to create class labels for the test vector. This approach will be demonstrated on a protein structure prediction problem in Chapter 4. Like its crisp counterpart, the fuzzy k-NN (FKNN) algorithm is simple in concept. Let $\mathbf{x}_1, \mathbf{x}_2, \ldots, \mathbf{x}_k$ be the k nearest neighbors of a test vector $\mathbf{x}$. The goal is to compute the membership of $\mathbf{x}$ in each class. The formula for the membership of $\mathbf{x}$ in the ith class, $u_i(\mathbf{x})$, as given in [Keller *et al.*, 1985] is

$$u_i(\mathbf{x}) = \frac{\sum_{j=1}^{k} u_{ij}\left(\dfrac{1}{\left(d(\mathbf{x}, \mathbf{x}_j)^{2/m-1}\right)}\right)}{\sum_{j=1}^{k}\left(\dfrac{1}{\left(d(\mathbf{x}, \mathbf{x}_j)^{2/m-1}\right)}\right)} \qquad (2.14)$$

where $m > 1$ is a constant (like the fuzzifier in the FCM). Hence, $u_i(\mathbf{x})$ is proportional to a weighted average of the inverse distances of $\mathbf{x}$ to each of its k nearest neighbors. The weights correspond to fuzzy i^{th} class labels of neighbors. The denominator is used to scale the memberships for all classes so that they sum to 1. In the crisp case, this is an inverse distance weighted k-NN [Dasarathy, 1991]. As with fuzzy clustering, if it is necessary to compute a crisp label, $\mathbf{x}$ can be assigned to the class with the largest membership.

Example 2.7. Consider the situation in Figure 2.23, the star representing a point x along with the 6 nearest neighbors, three from class 1 (circles) and three from class 2 (squares). Clearly, the crisp 6-NN would result in a tie. We are using an even number of neighbors in this example only to drive home this fact.

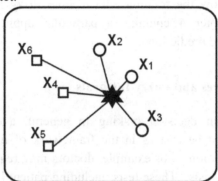

Figure 2.23 A point $\mathbf{x}$ (the star) with 6 neighbors split between class 1 (circles) and class 2 (squares).

Suppose now that the distances between $\mathbf{x}$ and these 6 vectors are ¼, ½, ½, ½, 1, and 1, respectively. If the class memberships are crisp, i.e., 1 in the designated class and 0 in the other, then for $m = 2$, Equation 2.14 produces

$$u_1(\mathbf{x}) = \frac{4+2+2}{4+2+2+2+1+1} = \frac{2}{3}, \text{ and}$$

$$u_2(\mathbf{x}) = \frac{2+1+1}{4+2+2+2+1+1} = \frac{1}{3}.$$

Hence, class 1 is favored, and so, if crisp assignments are required, the star would become a circle. However, suppose that x_1, x_2, and x_3 were not very typical members of class 1, all having class 1 memberships of 5/8 for simplicity (memberships in class 2 of 3/8) while the other three remain strongly in class 2. Then the calculations yield

$$u_1(\mathbf{x}) = \frac{(4+2+2)(5/8)}{4+2+2+2+1+1} = \frac{5}{12}, \text{ and}$$

$$u_2(\mathbf{x}) = \frac{(4+2+2)(3/8)+2+1+1}{4+2+2+2+1+1} = \frac{7}{12}.$$

Now, the Fuzzy 6-NN algorithm generates membership values for $\mathbf{x}$ that places it more strongly in class 2 due to the weaker confidence in the typicality of the class 1 points. A crisp partition would thus make the star a square. How the training data receive fuzzy labels is problem dependent. Chapter 4 contains a particular application to protein secondary structure prediction.

2.8 Fuzzy Measures and Fuzzy Integrals

Many problems in decision making in general, and specifically in bioinformatics, can be cast as in the framework of fusion of multiple sources of information. For example, doctors may request several tests to arrive at a diagnosis. These tests, including patient history, all supply partial evidence for possibly more than one conclusion. An expert diagnostician combines the results of the tests with "worth" of the individual assays, as well as groups of them, to support or refute particular conclusions. Some tests taken individually may only provide limited confidence in a decision, but taken as a group greatly increase that support. One of the advantages of fuzzy set theory is the wide range of computational mechanisms to implement such fusion of information. In this section we develop one of these powerful frameworks, the fuzzy integral. It is used to combine partial (objective) support for a hypothesis from the standpoint of individual sources of information together with (possibly subjective) weights of various subsets of these sources of information.

2.8.1 Fuzzy measures

The fuzzy integral is based on the concept of fuzzy measures, generalizations of probability measures, which in themselves will be shown to be effective to combine information in certain applications (see Chapter 3). Consider a finite set $X = \{x_1, x_2, ..., x_n\}$ of sources of information. Each x_i can be a diagnostic test, the expression level of a certain gene, an annotation term for a gene, etc. While only finite sets are considered here, the theory of fuzzy measures and fuzzy integrals can be extended to infinite sets (see [Grabisch *et al.*, 2000; Wang and Klir, 1992]).

Let 2^X denote the power set of X, i.e., the set of all subsets of X. A fuzzy measure, g, is a real valued function $g : 2^X \rightarrow [0,1]$, satisfying the following properties:

$$1. \ g(\varnothing) = 0 \ \text{and} \ g(X) = 1$$

$$2. \ g(A) \le g(B) \ \text{if} \ A \subseteq B$$

Note that the normal additivity condition of probability theory is replaced by the weaker condition of monotonicity (property 2). For a fuzzy measure g, let $g^i = g(\{x_i\})$. The mapping $x_i \rightarrow g^i$ is called a fuzzy density function. The fuzzy density value, g^i, is interpreted as the (possibly subjective) importance of the single information source x_i in determining the answer to a particular question, perhaps the similarity of two genes. Fuzzy measures are quite general since they only require two simple properties to be satisfied. However, it is often the case that the densities can be extracted from the problem domain or supplied by experts. The key to using fuzzy measures involves finding ones that can be built out of the densities. One of the most useful classes of fuzzy measures is due to Sugeno [Sugeno, 1977]. A fuzzy measure g is called a Sugeno measure (g_λ-fuzzy measure) if it additionally satisfies the following property [Sugeno, 1977]:

$$3. \ \text{For all} \ \ A, B \subseteq X \ \text{with} \ A \cap B = \varnothing,$$

$$g_\lambda(A \cup B) = g_\lambda(A) + g_\lambda(B) + \lambda g_\lambda(A)g_\lambda(B) \ \text{for some} \ \lambda > -1 \qquad (2.15)$$

Unless needed, the subscript λ will be omitted for simplicity. If the densities are known, the value of λ for any Sugeno fuzzy measure can be uniquely determined for a finite set X using Equation 2.15 and the facts $X = \bigcup_{i=1}^{n} \{x_i\}$ and $g_\lambda(X) = 1$, which leads to solving the following equation for λ:

$$(1+\lambda) = \prod_{i=1}^{n}(1+\lambda g^i) \qquad (2.16)$$

This equation has a unique solution for $\lambda > -1$ [Sugeno, 1977].

Example 2.8. To illustrate the calculation of a Sugeno fuzzy measure, suppose $X = \{x_1, x_2, x_3\}$ and suppose that $g^1 = 0.2$, $g^2 = 0.3$, and $g^3 = 0.1$ (we will see in Chapter 3 how such densities can be found in a bioinformatics application). Note that the resulting measure in this case cannot be a probability measure because the densities, i.e., the measures of the singleton subsets, do not add up to 1. Then, Equation 2.16 becomes

$$1 + \lambda = (1 + 0.2\lambda)(1 + 0.3\lambda)(1 + 0.1\lambda)$$

Expanding and collecting terms, λ must be the solution of the quadratic equation $0.006\lambda^2 + 0.11\lambda - 0.4 = 0$. While there are 2 solutions, only one of them, $\lambda = 3.2$, is greater than -1, as guaranteed by the theory. Hence, the complete fuzzy measure is shown in Table 2.8.

Table 2.8 Sugeno fuzzy measure for $X = \{x_1, x_2, x_3\}$ with densities 0.2, 0.3, and 0.1

Subset	Measure
ϕ	0.0
$\{x_1\}$	0.2
$\{x_2\}$	0.3
$\{x_3\}$	0.1
$\{x_1, x_2\}$	0.2+0.3+3.2(0.2)(0.3) = 0.69
$\{x_1, x_3\}$	0.2+0.1+3.2(0.2)(0.1) = 0.36
$\{x_2, x_3\}$	0.3+0.1+3.2(0.3)(0.1) = 0.5
$X = \{x_1, x_2, x_3\}$	0.36+0.2+3.2(0.36)(0.2) = 1.006 $\approx$ 1.0
	(we rounded the intermediate results)

In Chapter 3 we describe a method to build a similarity index between proteins using fuzzy measures.

2.8.2 Fuzzy integrals

Let X be a set and let $h: X \to [0,1]$ be a function that provides the support of a given hypothesis from the standpoint of each source of information, called a partial evaluation function. Suppose $g: 2^X \to [0,1]$ is a fuzzy measure. Then the Sugeno fuzzy integral is defined by

$$\int h(x) \circ g = \sup_{E \subseteq X} \left[\min(\min_{x \in E} h(x), g(E)) \right] = \sup_{\alpha \in [0,1]} \left[\min(\alpha, g(A_\alpha)) \right]$$

where $A_\alpha = \{x \mid h(x) \geq \alpha\}$.

For the finite case (the case that this chapter really addresses), suppose $h(x_1) \geq \cdots \geq h(x_n)$ (If this is not the case for any object instance, then reorder the set of information sources, X, so that this relation holds). Then the Sugeno fuzzy integral can be shown to be [Sugeno, 1977]

$$S_g(h) = \max_{i=1}^{n} \left[\min(h(x_i), g(A_i)) \right] \qquad (2.17)$$

where $A_i = \{x_1, \ldots, x_i\}$.

The reader is referred to [Grabisch *et al.*, 2000; Wang and Klir, 1992] for an extensive theoretical background on fuzzy measures and fuzzy integrals.

The original definition given by Sugeno for the fuzzy integral is not a proper extension of the Lebesgue integral, i.e., the integral from Calculus, in the sense that the Lebesgue integral is not recovered when the measure is additive. To avoid this drawback, Murofushi and Sugeno [Murofushi and Sugeno, 1991] proposed the Choquet fuzzy integral, referring to a functional defined by Choquet in a different context. Let h be the partial evaluation function on X with values in [0, 1] and g be a fuzzy measure. The Choquet integral is:

$$\int_X h(x) \circ g = \int_0^1 g(A_\alpha) d\alpha \quad \text{where } A_\alpha = \{x \mid h(x) \geq \alpha\}.$$

If X is a discrete set, the Choquet integral can be computed as follows:

$$C_g(h) = \sum_{i=1}^{n} [h(x_i) - h(x_{i+1})] \cdot g(A_i) \tag{2.18}$$

where $h(x_1) \geq \cdots \geq h(x_n)$, $h(x_{n+1}) = 0$, and $A_i = \{x_1, \ldots, x_i\}$.

It is also informative to write the discrete Choquet integral as a (nonlinear) weighted sum of these values in which the weights depend on their order. Define

$$\delta_i(g) = g(A_i) - g(A_{i-1}) \text{ for } i = 1, 2, \ldots, n \tag{2.19}$$

where we take $g(A_0)$ to be 0. Then

$$C_g(h) = \sum_{i=1}^{n} \delta_i(g) \cdot h(x_i) \tag{2.20}$$

Note that, in the general case, the sum in Equation 2.20 is a nonlinear combinations of the values of h because the ordering of the arguments $x_1, \ldots, x_n$ depends upon the relative sizes of the values of the function h.

Example 2.9. As a simple illustration, suppose that the set of information sources is $X = \{x_1, x_2, x_3\}$ with the Sugeno fuzzy measure specified in Table 2.8. For the partial evaluation function $h: X \to [0,1]$, given by $h(x_1) = 0.9$, $h(x_2) = 0.7$, and $h(x_3) = 0.2$, we calculate the Sugeno and Choquet fuzzy integrals. First note that the function values are already sorted in descending order so that there is no need to re-order the set of information sources. From Equation 2.17, the Sugeno fuzzy integral of h with respect to the fuzzy measure is given by

$$S_g(h) = (0.9 \wedge 0.2) \vee (0.7 \wedge 0.69) \vee (0.1 \wedge 1.0) = 0.69.$$

Similarly, Equation 2.18 produces a Choquet fuzzy integral of

$$C_g(h) = (0.9 - 0.7)(0.2) + (0.7 - 0.2)(0.69) + (0.2 - 0.0)(1.0) = 0.59.$$

As will be seen in Chapter 3, for particular fuzzy measures, g, the Choquet fuzzy integral reduces to a linear combination of order statistics. If the weights (thought of as a fuzzy subset of the positive integers from 1 to n) have a linguistic interpretation, this integral has become known as an Ordered Weighted Average (OWA) [Yager, 1988; Yager, 1993]. The measures that produce these special operators are very specific in that they must be constant on all subsets of information sources that contain the same number of elements. Hence, the full Choquet (and Sugeno) fuzzy integrals represent a very broad class of information fusion mechanisms that can be tailored to the problem at hand.

The key to using fuzzy integrals to fuse multiple sources of information is to construct the fuzzy measures that specify the worth of all subsets of sources of information. In Chapter 3, Sugeno measures are employed. As noted above for these fuzzy measures, only densities (the worth of each singleton information source) need be specified. This is problem dependent and we will provide examples in Chapter 3. There are numerous methods to automatically learn either densities or entire measures if training data are available [Grabisch *et al.*, 2000; Keller *et al.*, 2000; Tahani and Keller, 1990; Keller *et al.*, 1994; Mendez-Vazquez *et al.*, 2007]. These techniques dramatically increase the applicability of fuzzy integrals for general information fusion.

2.9 Summary and Final Thoughts

We began this chapter with a motivation as to why fuzzy models should be considered in bioinformatics. With that motivation, we introduced some of the fundamental concepts of fuzzy set theory to form a basis for the applications to bioinformatics. We defined and showed examples of membership functions and some of the many types of aggregation operators available to us to fuse partial confidences into a global confidence estimate. After grounding ourselves in the underpinnings, we successively described fuzzy relations and fuzzy logic inference, fuzzy clustering, the fuzzy k-nearest neighbors, fuzzy measures, and fuzzy integrals. These topics were selected both to provide a sampling of the breath of fuzzy models and to provide the notation and mathematics for applications covered in subsequent chapters of this book. We tried to give the readers an appreciation for the beauty and the utility of fuzzy set theory and fuzzy logic. Clearly, we have only scratched the surface of fuzzy models and hope that our initiation engenders a desire for further study.

Chapter 3

Fuzzy Similarities in Ontologies

3.1 Introduction

In philosophy, ontology is a branch of metaphysics that studies "the reasoning (logos) about being (ontos)" [http://en.wikipedia.org/Ontology]. As opposed to this theoretical meaning, in computer science, an ontology is a pragmatic representation of a particular area of information. Specifically, an ontology consists of a set of concepts, regarding the area of information, and the relations between them. Hence, this pragmatic representation can be used to facilitate knowledge sharing, retrieval, and discovery in the area represented by the ontology. An ontological representation serves two purposes: (1) it is used to create controlled vocabularies, and (2) it encodes common knowledge of the area.

Among the best known bio-ontologies are: RiboWeb [http://riboweb.stanford.edu/riboweb/], EcoCyc [http://ecocyc.org], the ontology for molecular biology (MBO) [Schulze-Kremer 1998], the Gene Ontology (GO) [http://www.geneontology.org], TAMBIS (TaO) [Baker *et al.*, 1998] and the KEGG Ontology (KO) [http://www.genome.jp/kegg]. In addition, the MeSH ontology [http://www.nlm.nih.gov/mesh/] is employed in searching for biomedical publications and gene product annotation [Perez *et al.*, 2004].

Compared to other bio-ontologies, the GO has a narrower scope [Stevens *et al.*, 2000], which centers on the role of a gene product in an organism instead of describing molecular biology as a whole. However, the GO has emerged as the ontology of choice in many bioinformatics applications such as automatic annotation of gene function [Khan *et al.*, 2003; Martin *et al.*, 2004], microarray analysis [Khatri *et al.*, 2004; Al Shahrour *et al.*, 2004], and gene family clustering [Speer *et al.*, 2004; Pal *et al.*, 2005]. Since all the examples from this chapter will be based on GO, we will describe it in more detail in what follows.

The Gene Ontology has three branches (hierarchies) (see Figure 3.1).

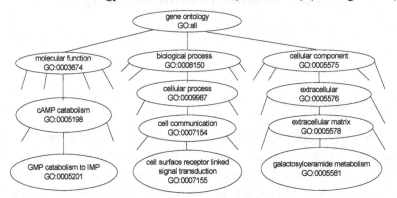

Figure 3.1 Partial view of the Gene Ontology (GO) hierarchy.

The first branch, "molecular function", is centered on the function of a gene product. The second one, "biological process", is related to the processes in which a gene product may be involved. The third branch, "cellular component", describes cellular location and structure. In the GO, the concepts are linked by two types of relations: "is-a" and "part-of". From similarity stand point, it is obvious that the two types of relations are quite different. While all the descendents of a term are quite similar in the "is-a" case, they are usually quite dissimilar for the "part-of" relation. In this chapter we ignore the "part-of" case and we will consider that all the relations between terms are of type "is-a". The GO contains a high level of detail [Stevens, 2000] and is updated on a daily basis. As of June 2007, it has about 23,200 concepts.

Ontologies have been used as controlled vocabularies for knowledge sharing and retrieval in many areas. As opposed to plain text, the GO is more effective in indexing gene products, resulting in improved database searches. For example, when searching for genes involved in cell death using the GO term "cell apoptosis" (GO identifier GO:0006915), we are able to retrieve genes involved in the apoptosis process such as BCL2 and RAF1, that could not have been related using a regular (syntactical) database search. The advent of the world-wide web increased the need for knowledge sharing and interoperability. In this direction, the Web Ontology Language (OWL) [http://www.w3.org/2004/OWL/] aims at facilitating the communication (information exchange) between various web agents.

The knowledge encoded in an ontology has been used in computer sciences such as artificial intelligence (AI) and computational intelligence (CI). The distinction that we make here between AI and CI is (see Chapter 2, Section 2.1) that while the former uses primarily symbolic algorithms, the latter employs numerical algorithms. In AI, an ontology is viewed as a set of symbolic rules in the form "A is-a B" (e.g., "a protein tyrosine phosphatase is-a protein phosphatase" [Wolstencroft *et al.*, 2005]) or "A part-of B" ("ATP binding motif is part-of ABC transporter"). In some approaches, ontologies are related to predicate logic and are usually implemented using symbolic languages such as LISP or PROLOG [Horrocks *et al.*, 1998]. In other approaches, ontologies are developed using description logic (DL) and the inferences are performed by dedicated reasoners. This approach has been used, for example, to classify protein phosphatases [Wolstencroft *et al.*, 2005]. Although rigorous, this technique has difficulty in dealing with imprecise knowledge. As we have seen in Chapter 1, there are various biological problems where a fuzzy logic approach may be more suitable than a traditional approach. In this chapter we will show several situations where fuzzy logic can be used in conjunction with ontologies.

There are three aspects of the relation between fuzzy logic and ontologies from a CI perspective. First, we can apply known fuzzy logic algorithms such as fuzzy clustering [Pal *et al.*, 2005] and fuzzy measure theory [Popescu *et al.*, 2006] to objects described by ontology concepts. Second, an ontology may be used to define fuzzy matches between concepts based on their similarity. As a consequence, we can use fuzzy matches to retrieve information from databases [Andreasen *et al.*, 2003] or to build fuzzy rule systems where the rules are fired ontologically instead of using membership functions [Popescu *et al.*, 2007]. Third, an ontology itself may be fuzzy. Fuzzy ontologies have been used in text mining [Tho *et al.*, 2006] and text summarization [Lee *et al.*, 2005]. In a fuzzy ontology an object may have different degrees of memberships in various classes (concepts). This approach is usually required for numeric data where numbers have to be mapped to concepts (such as 198 cm to concept "tall"). In this chapter we do not discuss fuzzy ontologies.

In the CI approach, the classification and knowledge discovery algorithms are based on distance between objects. Usually, the distance is calculated from a set of numeric attributes (features) of the objects such as molecular weight or number of nucleotides in the sequence. However, this distance computation becomes more complicated when the object attributes are symbolic (not numeric). Here we assume that the objects, in our case gene products, are described by concepts from an ontology. The key to this CI approach is the quantification of the similarity between concepts based on their relation in a common ontology. Since distance d and similarity s are dual concepts we will use them interchangeably. Once the similarity between two concepts is evaluated to a number (between 0 - dissimilar, and 1- identical) then a variety of algorithms can be applied.

In the rest of the chapter we will define a number of fuzzy similarity measures and describe several ontological data mining algorithms and their applications in bioinformatics. The data mining algorithms presented here are intended to be practical applications of the fuzzy logic concepts introduced in Chapter 2.

3.2 Definition of Ontology-Based Similarity

Similarity is a central concept in knowledge discovery. The relations between objects are quantified by their similarity. The computation of similarity depends on the representation of the objects. In our case, the objects (gene products) are represented (annotated) using GO terms. That is, a gene product G is represented as a set of N GO terms as $G=\{T_1,...,T_N\}$.

The computational use of the GO requires a definition of similarity s_{12} (or dissimilarity) between two terms T_1 and T_2 from the ontology. There are two main approaches to computing term similarity. One approach employs aspects of the topological arrangement of the terms in the ontology [Jiang, 1997] such as the path between concepts, the density of child-concepts of a node or the depth of the concepts in the hierarchy. Related topological approaches can be found in [Andreasen *et al.*, 2003] and [Joslyn *et al.*, 2004]. A simple path-based computation of term similarity is shown in Figure 3.2.

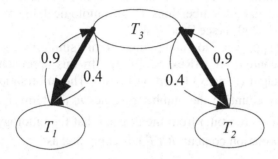

Figure 3.2 Example of path-based computation of the similarity (membership) between two GO terms. Note that the thin arcs represent the weight assignment process while the thick arcs represent the ontological relation "is-a"(GO is a directed acyclic graph). Here, the similarity between T_1 and T_2 is 0.9*0.4=0.36.

The similarity between two ontology terms T_1 and T_2, s_{12}, is computed as [Andreasen *et al.*, 2003]:

$$s_{12} = \max_{\{P_i\}} \prod_{j \in P_i} w_{ij}, \qquad (3.1)$$

$\{P_i\}$ is the set of all possible paths connecting T_1 and T_2 in the GO and w_{ij} is the weight assigned to the arc j from path P_i. In the example from Figure 3.2, the similarity between T_1 and T_2 is $s_{12} = 0.9 * 0.4 = 0.36$. Here, we consider only two types of weights: specialization weights (upward from the ancestor node to the descendent node) with a value of 0.9 and generalization weights (downward from the descendent to the ancestor) with a value of 0.4. The weights were assigned somewhat arbitrary here, but they can be computed based on term co-occurrence data [Lau, 2007]. We note that the weights themselves have an opposite direction from the process they represent. For instance, although the specialization relation runs downward (thick arrows in Figure 3.2), the specialization weights (with value 0.9) run upward. The difference between generalization ("serine-threonine kinases are kinases") and specialization ("kinases are serine-threonine kinases") is that while the first holds entirely (high membership value, arbitrarily chosen as 0.9 in Figure 3.2) the latter holds only partially ("medium" membership value, arbitrarily chosen as 0.4 in our case). As can be observed from Equation 3.1, the above measure is not symmetric. Hence, it is not, strictly speaking, a similarity. However, this approach was found to be useful in modeling the fuzzy memberships in an ontological fuzzy rule system [Popescu *et al.*, 2007] (see Section 3.8).

Another approach to computing term similarity is based on information content (IC) [Resnik, 1999]. In this approach, each term receives a weight (density) based on its IC. The information content is determined by counting the number of times n_i the term T_i appears in a set of documents (corpus) from the domain that the ontology represents. Then the information content, $IC(T_i)$, is computed as:

$$IC(T_i) = \frac{\log_2 \left(\dfrac{\sum_{j=1}^{N_i} n_j}{\sum_{j=1}^{N} n_j} \right)}{\log_2 \left(\dfrac{1}{\sum_{j=1}^{N} n_j} \right)}, \qquad (3.2)$$

where N_i is the number of ontology nodes that have node i as ancestor and N is the total number of terms from the ontology. We note that in contrast to previous definitions of IC [Resnik, 1999], the one above produces values in $[0,1]$. As a consequence, the root node (term denoted "GO:all" in Figure 3.1) has $IC=0$ and any term i that is a leaf node (does not have any child nodes) and appears just once in the corpus (that is, $n_i=1$) has $IC=1$. In addition, the information content of a child node is greater than that of any of its parent nodes.

Employing the term information content, the similarity s_{12} between two ontology terms T_1 and T_2 may be defined as [Resnik, 1999]:

$$s_{12}(T_1,T_2) = IC(NCA\{T_1,T_2\}),\qquad(3.3)$$

that is, the information content of the nearest common ancestor (NCA), of the two terms in the ontology tree. The disadvantage of the above definition is that in some cases it produces inconsistent results (see Figure 3.3).

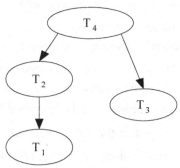

Figure 3.3 Example of inconsistent result of the information content method as defined in [Resnick 1999]. In this case, $s_{13}(T_1,T_3)=s_{23}(T_2,T_3)$ since they have the same NCA, T_4. However, intuitively, s_{13} should be smaller than s_{23}.

A possible solution to the above problem [Jiang, 1997] consists of a combined topological and information-content approach. In this approach, the weight w_{ij} between the term T_i and term T_j, is calculated as the difference of their information content $IC(T_i)-IC(T_j)$. Another solution [Lin, 1998] is to normalize s_{12} from Equation 3.3 using the information content of the two terms T_1 and T_2 as

$$s_{12}(T_1,T_2) = \frac{2IC(NCA\{T_1,T_2\})}{IC(T_1)+IC(T_2)},\qquad(3.4)$$

where $IC(T_1)$ and $IC(T_2)$ are the IC of the two terms and $IC(NCA\{T_1,T_2\})$ is the IC of their nearest common ancestor. It can be seen that for this case, in Figure 3.3, $s_{13} < s_{23}$ since $IC(T_2) < IC(T_1)$ due to the definition of IC (Equation 3.3).

3.3 Set-Based Similarity Measure

Consider two gene products, G_1 and G_2, represented by collections of concepts (annotations) from an ontology O as $G_1 = \{T_{11},...,T_{1j},...,T_{1n}\}$ and $G_2 = \{T_{21},...,T_{2j},...,T_{2m}\}$, $T_{ij} \in O$, $i \in \{1,2\}$. In addition, suppose each term $T_{ij} \in O$ from the two sets is assigned a weight $g^{ij} \in [0,1]$ that in our case is its information content, IC.

Our goal is to define a similarity measure between G_1 and G_2, $s(G_1,G_2)$. The similarity, s, can be computed using a set or a vector space approach. Let N be the number of concepts in O. In the vector space approach, each gene product G_i is described by a vector $\mathbf{v}_i \in R^N$, where v_{ij} is 1 if T_j is present in the G_i representation and 0 else. If the number of concepts in the ontology O is large ($N >> 0$) and the cardinality of the annotation set (m and n) is small, the vectors $\mathbf{v}_i$ become long and sparse, making subsequent use of the vectors in algorithms like clustering problematic. This is especially true in the case of gene products annotation where the typical annotation contains less than 10 concepts while the GO (in June 2007) has 23,200 terms.

An alternative way to compute the similarity is to consider a set approach in which each gene product is represented by a set of terms as mentioned above. There are two main set approaches to computing the similarity s: a "pair-wise" aggregation and the "bag of words".

3.3.1 Pair-wise aggregation

In the "pair-wise" approach, the similarity s is computed by aggregating the ontology pair-wise terms similarities, $s_{ij}(T_{1i},T_{2j})$, for the pairs of all terms from G_1 and G_2 respectively. For the sake of notations, we relabel the values $\{s_{ij} \mid i=1,...,n, j=1,...,m\}$ as $\{S_1,...,S_k,...,S_{mn}\}$. In [Pal *et al.*, 2005], a generalized pair-wise similarity was defined using Ordered Weighted Average (OWA) [Yager 1996] operators as:

$$s_{OWA}(G_1,G_2) = \sum_{k=1}^{nm} w_k S_{(k)}, \qquad (3.5)$$

where $S_{(k)}$ denotes a reordering of the pair-wise term similarities S_k such that $S_{(1)} \geq S_{(2)} \ldots \geq S_{(mn)}$, and **w** is a weight vector with $\sum_{k=1}^{nm} w_k = 1$. From the above formula we obtain the average pair-wise similarity [Lord *et al.*, 2003] by taking $w_i = 1/nm$, $i = 1, \ldots, nm$. Similarly, the maximum pair-wise similarity is obtained by setting $w_1 = 1$ and $w_i = 0$, $i = 2, \ldots, nm$. The OWA similarity requires the following normalization:

$$s(G_1, G_2) = \frac{s_{OWA}(G_1, G_2)}{\max\{s_{OWA}(G_1, G_1), s_{OWA}(G_2, G_2)\}}. \tag{3.6}$$

Without normalization, the average is not a true similarity (since $s(G_1, G_1) < 1$). OWA aggregation is very general, providing all linear combinations of order statistics. Pal *at al.* [2005] proved that this normalization produced valid similarity numbers for a subset of all possible weight vectors. We note that the weights for the average ($w_i = 1/nm$) were not among the subset from Pal's proof. However, in our experience, normalizing the average using Equation 3.6 produces consistent similarity values. The maximum pair-wise similarity tends to overestimate the similarity $s(G_1, G_1)$ since it is enough that the two gene products share one GO term for the similarity to be 1. This is especially bad for a multi-domain protein. If two multi-domain protein share functions (hence GO terms), their similarity will be 1, making impossible any discrimination among them. The main drawback of the pair-wise approach is slow computational speed due to consideration of the all *nm* combinations of concepts.

Example 3.1. Consider two human anti-apoptotic genes BAG1 and BCLW, both members of the BCL2 gene family. Since the two genes are from the same family and have a common function we expect that they exhibit some reasonable degree of similarity. However, the sequence similarity (computed using the Smith-Waterman algorithm) in the case of the two genes is about 0.03 and BLAST does not detect any significant similarity. Hence, using just the sequence similarity we would say that the two genes are not related.

The GO annotations for BAG1 (four terms) and BCLW (three terms) are shown in Tables 3.1 and 3.2, respectively. In addition to the ID of each GO term, we show its type (i.e. GO branch: "M" for "molecular function", "P" for "biological process", "C" for "cellular component", see Figure 3.1), its evidence type (the way the annotation was performed: TAS-traceable author statement, IPI-inferred from physical interaction) and its calculated IC (using Equation 3.2).

Table 3.1 GO annotation set for BAG1_HUMAN

Name	receptor signaling protein activity	cytoplasm	anti-apoptosis	cell surface receptor linked signal transduction
Term ID	GO:0005057	GO:0005737	GO:0006916	GO:0007166
Evidence	TAS	TAS	TAS	TAS
Type	F	C	P	P
IC	0.49	0.10	0.52	0.29

Table 3.2 GO annotation set for BCLW_HUMAN

Name	protein binding	anti-apoptosis	spermatogenesis
Term ID	GO:0005515	GO:0006916	GO:0007283
Evidence	IPI	TAS	TAS
Type	F	P	P
IC	0.22	0.52	0.49

The pair-wise similarity for the GO terms from the two above annotation sets is given in Table 3.3, Table 3.4 and Table 3.5.

Table 3.3 Pair-wise similarity for the GO terms from the BAG1 and BCLW annotation sets (used in numerator of Equation 3.6)

BAG1 ⇓, BCLW⇒	GO:0005515	GO:0006916	GO:0007283
GO:0005057	0.35	0	0
GO:0005737	0	0	0
GO:0006916	0	1	0.05
GO:0007166	0	0.25	0.28

Table 3.4 Pair-wise similarity for the GO terms from the BCLW annotation set (used in the denominator of Equation 3.6)

	GO:0005515	GO:0006916	GO:0007283
GO:0005515	1	0	0
GO:0006916	0	1	0.05
GO:0007283	0	0.05	1

Table 3.5 Pair-wise similarity for the GO terms from the BAG1 annotation set (used in the denominator of Equation 3.6)

	GO:0005057	GO:0005737	GO:0006916	GO:0007166
GO:0005057	1	0	0	0
GO:0005737	0	1	0	0
GO:0006916	0	0	1	0.25
GO:0007166	0	0	0.25	1

Since BCLW and BAG1 share one annotation term (GO:0006916), the maximum pair-wise similarity, $s_{Max}=1$. The average similarity is:

$$s_{Ave} = \frac{s_{OWA}(BAG1, BCLW)}{\max\{s_{OWA}(BAG1, BAG1), s_{OWA}(BCLW, BCLW)\}}$$

$$= \frac{0.16}{\max\{0.28, 0.34\}} = 0.47$$

where, for example, $s_{OWA}(BAG1, BCLW)$ is calculated using Equation 3.5 with a weight vector $\{w_i=1/12\}_{i=1,12}$. As mentioned above, s_{max} overestimates the similarity between the two genes while s_{ave} is underestimating it by considering all the combinations of terms. To obtain an intermediate value between the two we can use, for instance, an "al least 2" OWA operator that has the weights $w_1=0.5$, $w_2=0.5$ and $\{w_i=0\}_{i=3,12}$. In this case the similarity becomes:

$$s_{At\_least\_2} = \frac{0.5*1+0.5*0.35}{\max\{0.5*1+0.5*1, 0.5*1+0.5*1\}} = \frac{0.67}{1} = 0.67.$$

3.3.2 Bag of words similarities

In the "bag of words" approach, the similarity measure between two gene products G_1 and G_2 considers the two sets of concepts directly. Various "bag of words" methods have been described in the literature [Manning *et al.*, 2001]. Here, we mention a few of them:
- Jaccard similarity:

$$s_J(G_1,G_2) = \frac{|G_1 \cap G_2|}{|G_1 \cup G_2|};\qquad(3.7)$$

- Set cosine similarity:

$$s_C(G_1,G_2) = \frac{|G_1 \cap G_2|}{\sqrt{|G_1\| G_2|}};\qquad(3.8)$$

- Dice similarity:

$$s_D(G_1,G_2) = \frac{2|G_1 \cap G_2|}{|G_1|+|G_2|}.\qquad(3.9)$$

In the above formulas, $|G|$ denotes the cardinality of the set G, while $\cap$ and $\cup$ are standard set intersection and union. Although the "bag of words" similarities are faster than the pair-wise ones, they have the disadvantages that they do not consider the term relatedness and importance (weight). While they can be fuzzified to include the importance (information content) of the terms, they can not account for their semantic content (the term meaning, i.e. the relations between terms). For instance, the fuzzy Jaccard similarity is defined by:

$$s_{FJ}(G_1,G_2) = \frac{\displaystyle\sum_{\{i|T_i \in G_1 \cap G_2\}} IC(T_i)}{\displaystyle\sum_{\{j|T_j \in G_1 \cup G_2\}} IC(T_j)},\qquad(3.10)$$

where $IC(T_i)$ represents the information content of term T_i.

Example 3.2. For the same two genes from Example 3.1, BAG1 and BCLW, the above similarities are: $s_J = 0.17$, $s_C = 0.29$ and $s_D = 0.28$. The fuzzy Jaccard is $s_{FJ} = 0.25$ (using the information content row, IC, from Table 3.1 and 3.2). All the above values are essentially similar, indicating that the two genes are somewhat related but at fairly low level. This is true, since both genes are in the same family (BCL2) and they both have a anti-apoptotic (against death) role in the cell. The fuzzy Jaccard similarity is 50% higher than the crisp one indicating that the common function (anti-apoptosis) of the two genes is annotated by a term that has a somewhat high IC (0.52).

The greatest disadvantage of the bag similarities is that they do not consider the relations between terms. As a consequence, these similarities are not informative when the number of terms in the annotation set (*m* and *n*, respectively) is small and the ontology is large. In this case, there is a high chance of having empty intersection between the two annotation sets, that is $G_1 \cap G_2 = \varnothing$, resulting in a zero similarity. In addition, even in the fuzzified case, the context of the term (how important a term is as compared to the others in the annotation set) is not considered since a term could have a different contribution to similarity depending on the annotation set.

To address two of the above problems (IC and context) in [Popescu *et al.*, 2006] we proposed a fuzzy measure similarity (FSM). FMS is able to account for information content and to consider the term context, as illustrated in the following section. Furthermore, the augmented FMS (AFMS) avoids the zero similarity problem by considering the relations between terms as given by the related ontology (the GO in our case). We note that the augmented fuzzy Jaccaard could also address the IC and the zero symilairty problems but not the context one.

3.4 Fuzzy Measure Similarity

The fuzzy measure similarity (FMS) between two sets, G_1 and G_2, of ontology terms is defined as [Popescu *et al.*, 2006]:

$$s_{FMS}(G_1, G_2) = \frac{g_1(G_1 \cap G_2) + g_2(G_1 \cap G_2)}{2}, \qquad (3.11)$$

where g_1 and g_2 are Sugeno measures (see Chapter 2, Section 2.7.1) defined on G_1 and G_2 respectively. Here, the fuzzy densities that determine the measures are given by the information content of the terms, i.e., $\{g^{1i}\} = IC(T_{1i})_{i=1,...,n}$ and $\{g^{2j}\} = IC(T_{2j})_{j=1,...,m}$.

Example 3.3. Consider the same genes from example 3.1, BAG1 and BCLW. To make the calculation less trivial we replace the term GO:0007283 by the term GO:0005057 in the BCLW annotation set. Now, the two genes share two terms, hence their Jaccard similarity is 0.4 indicating a medium-low level of similarity. The Sugeno measure parameters for the densities associated with the genes are calculated by solving $(\lambda-0.49)(\lambda-0.1)(\lambda-0.52)(\lambda-0.29)=1-\lambda$, resulting in $\lambda_1=-0.67$ and $(\lambda-0.22)(\lambda-0.52)(\lambda-0.49)=1-\lambda$, resulting in $\lambda_2=-0.52$, respectively. The measure of the intersection in the BAG1 context is $g_1(IC(GO:0006916,GO:0007283))=$ IC(GO:0006916) + IC(GO:0007283)$+\lambda_1*$IC(GO:0006916)*IC(GO:0007283)=0.52+0.49-0.67*0.52*0.49=0.83, while in the BCLW context is $g_2(IC(GO:0006916,GO:0007283))=0.52+0.49-0.52*0.52*0.49=0.88$. While the two fuzzy measures differ by only 5% here, it is apparent that they could be very different if, for instance, a term with very high IC would be present in one of the annotation sets. Finally, the FMS in this case is $s_{FMS}=0.5(0.83+0.88)=0.85$, indicating a high level of similarity.

Although FMS accounts for the term context, it has a similar problem to the "bag of words" methods: if the intersection of the sets is empty the measure is not informative (because the measure of the empty set is 0). To deal with this problem, we augment the initial annotation set to ensure that the intersection is non-empty.

3.5 Fuzzy Measure Similarity for Augmented Sets of Ontology Objects

The approach taken in [Popescu *et al.*, 2006] is to augment each annotation set with the nearest common ancestor (NCA) of every pair (T_{1i}, T_{2j}), called $T_{1i,2j}$. The augmented sets become:

$$G_1^+ = G_1 \cup \{T_{1i,2j}\} \text{ and } G_2^+ = G_2 \cup \{T_{1i,2j}\} \qquad (3.12)$$

and the resulting augmented intersection is:

$$[G_1 \cap G_2]^+ = [G_1^+ \cap G_2^+] = [G_1 \cap G_2] \cup \{T_{1i,2j}\}. \qquad (3.13)$$

Using this new intersection, $[G_1 \cap G_2]^+$, the augmented FMS (AFMS), denoted by s_{AFMS}, is defined as:

$$s_{AFMS}(G_1, G_2) = \frac{g_1^+([G_1 \cap G_2]^+) + g_2^+([G_1 \cap G_2]^+)}{2}, \qquad (3.14)$$

where g_k^+ is the fuzzy measure computed on G_k^+, $k=1,2$. In fact, the intersection set will contain the NCA's of all term pair, but they will have different contributions to the similarity depending on the context of each annotation set.

3.6 Choquet Fuzzy Integral Similarity Measure

The Choquet integral (see Chapter 2, Section 2.7.2) is a nonlinear method of combining sources of information (annotation terms in our case) and the reliability of these information sources. The underlying hypothesis is that annotation uncertainty (reliability) should be included in the calculation of the similarity of the two gene products.

In the case of the Gene Ontology, the uncertainty is related to the source of the information that supports the annotation (see http://www. geneontology.org/GO.evidence.shtml). Several types of evidence are shown in Table 3.6. The GO annotations are assigned either manually by its curators based on published data, or automatically by specialized software based on various database information. Obviously, the annotations assigned based on published experiments, tagged as "Traceable Author Statement" (TAS), are more reliable than the ones inferred using automatic annotation software, labeled as "Inferred from Electronic Annotation" (IEA). In order to use the evidence types in the Choquet integral, we attach numeric confidence (reliability) values, c, to the evidence types. Here, the numeric confidence values are chosen somewhat arbitrarily, the only condition being that they obey the intuitive relations between annotation sources, such as $c_{TAS}>c_{IPI}>c_{IEA}>c_{ND}$. However, the Choquet integral framework allows for the computation of the above numeric values given training data in the form of target similarities for a set of gene products.

Table 3.6 Several types of evidence used in GO annotation

Traceable author statement	Inferred from sequence similarity	Inferred from electronic annotation	Inferred from pair-wise interaction	Not documented
TAS	ISS	IEA	IPI	ND

We want to compute the similarity of the same two gene products $G_1 = \{T_{11},...,T_{1i},...,T_{1n}\}$ and $G_2 = \{T_{21},...,T_{2j},...,T_{2m}\}$ taking into account the related confidence of the term evidence $c_1 = \{c_{11},...,c_{1i},...,c_{1n}\}$ and $c_2 = \{c_{21},...,c_{2j},...,c_{2m}\}$. The confidence (reliability) of a pair of terms, $c_{ij}(T_{1i},T_{2j})$, is computed as $c_{ij}(T_{1i},T_{2j}) = f(c_{1j},c_{2j})$, where f can be the maximum, average, or minimum operator and c_{1i} and c_{2j} are the confidences of assigning the annotations T_{1i} and T_{2j}, respectively. To simplify the notation we relabel $T_k = (T_{1i},T_{2j})$ for some i and j and its related confidence $c_k = c_{ij}(T_{1i},T_{2j})$ where $k=1,...,nm$. Then, the Choquet similarity is computed as follows:

$$s_{Cho}(G_1,G_2) = \sum_{i=1}^{nm} \left[s(T_{(i)}) - s(T_{(i+1)}) \right] \cdot g(S_i), \qquad (3.15)$$

where the pair-wise similarity values $s(T_k)$ are reordered so that $s(T_{(1)}) \geq s(T_{(2)}) \geq \cdots \geq s(T_{(nm)})$, $s(T_{(nm+1)}) = 0$, $S_i = \{T_{(1)}, \cdots, T_{(i)}\}$ and g is the fuzzy measure generated by the set of fuzzy densities $\{c_{ij}\}$. An intuitive representation of the Choquet integral is shown in Figure 3.4.

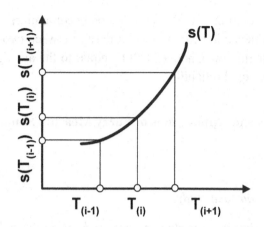

Figure 3.4 Intuitive representation of the Choquet integral. If $T_{(i)}$ were real numbers then the integral, the area under the curve $s(T)$ is calculated as $I = \sum [T_{(i)} - T_{(i-1)}] \cdot s(T_{(i)})$. In this case, the contribution of $T_{(i)}$ to $s(T_{(i)})$ is calculated as the measure of the intervals $[T_{(i-1)}, T_{(i)}]$, that is $T_{(i)}-T_{(i-1)}$. However, since $T_{(i)}$ represent pairs of ontology terms with associated uncertainties $c_{(i)}$, then one has to employ a fuzzy measure g to compute the contribution of $T_{(i)}$ as $g\left(\{T_{(1)}, \cdots, T_{(i)}\}\right) - g\left(\{T_{(1)}, \cdots, T_{(i-1)}\}\right)$. Then the integral can be written as $C = \sum_{i=1} [g(S_i) - g(S_{i-1})] \cdot s(T_{(i)})$, which is equivalent to Equation (3.15).

Example 3.4. For the same two genes from the previous examples, the ordered pair-wise term similarities (from Table 3.3) are $\{s(T_{(i)})\} = \{1, 0.35, 0.28, 0.25, 0.05, 0, 0, 0, 0, 0, 0, 0\}$. Considering (arbitrarily) $c_{TAS}=0.8$ and $c_{IPI}=0.5$, the corresponding pair-wise reliabilities (combined using product) are $\{c_{(i)}\} = \{0.64, 0.4, 0.64, 0.64, 0.64, 0.64, 0.64, 0.64, 0.64, 0.64, 0.4, 0.4\}$ resulting in $\lambda=-0.99$. Then, the Choquet similarity of BAG1 and BCLW1 is:

$$s_{Cho} = 0.64(1-0.35) + 0.78(0.35-0.28) + 0.92(0.28-0.25)$$
$$+ 0.97(0.0.25-0.05) + 0.99(0.0.5-0) = 0.74.$$

where, for example, $g(S_2)=0.64+0.40.99*0.64*0.4=0.78$.

This value (0.74) is smaller than that generated by FSM (0.85, see Example 3.3) because it includes in its calculations the reliability of the annotations. In addition, the value given by the Choquet similarity is between the average (0.47, underestimated) and maximum (1, overestimated) and depends on the confidence values assigned to the sources of annotation. As the annotations become more reliable (hence the term confidence increases), the similarity between two gene products increases. Finally, the Choquet will be equal to the average similarity if all terms have equal reliability.

3.7 Examples and Applications of Fuzzy Measure Similarity Using GO Terms

3.7.1 Lymphoma case study

In this section we introduce a case study used throughout this book that consists of 30 genes, most of them present on a methylation cDNA microarray constructed at the University of Missouri-Columbia to investigate lymphomas in humans. Lymphoma is a hematological neoplasm produced by mutated lymphocytes. Lymphocytes make up about a third of the white blood cells and are produced in the lymphatic tissue (such as lymph nodes, spleen, thymus and tonsils). These cells play an important role in the body immune system. Traditionally, lymphomas are classified as Hodgkin's and non-Hodgkin's. In fact, there are many types of lymphoma, each type being related to a given type of lymphocyte (helper T cells, killer T cells and B cells) and its developmental stage.

Table 3.7 Case study genes, denoted as GD30, used in lymphoma investigation.

No.	Gene Name	Gene Description	AA length	GeneBank Locus ID
1	RAF1	v-raf-1 murine leukemia viral oncogene homolog 1	274	NM_002880
2	ANXA1	annexin A1	193	NM_000700
3	B2L10	BCL2-like 10 (BCLL10)	175	NM_020396
4	BAG1	BCL2-associated athanogene	198	NM_004323
5	BCLW	BCL2-like apoptosis regulator (BCL2L2)	228	NM_004050
6	BFL1	BCL2-related protein A1, hemopoietic-specific early response (BCL2A1)	239	NM_004049
7	SOCS2	suppressor of cytokine signaling 2	166	NM_003877

8	BNIP1		195	NM_001205
9	BCL2	B-cell CLL/lymphoma 2	233	NM_000633
10	COF1	Cofilin (CFL1)	920	NM_005507
11	ASC	Apoptosis-associated speck-like protein	160	NM_013258
12	BCL10	B-cell CLL/lymphoma 10	198	NM_003921
13	BCLF1	Bcl-2-associated transcription factor 1 (BCLAF1)	277	NM_014739
14	BIK	Bcl-2-interacting killer (Apoptosis inducer NBK)	351	NM_001197
15	BIM	Bcl-2-like protein 11 (Bcl2-interacting mediator of cell death) (BCL2L11)	680	NM_006538
16	CASP3	caspase 3, apoptosis-related cysteine peptidase	281	NM_004346
17	CD2	CD2 antigen (p50), sheep red blood cell receptor	168	NM_001747
18	P73L	Tumor protein p73-like, oncogene	326	NM_003722
19	TNFSF10	Tumor necrosis factor ligand superfamily member 10	375	NM_003810
20	BAD	Bcl2-antagonist of cell death	313	NM_001901
21	FOSL2	Fos-related antigen 2	479	NM_005253
22	CD14	Monocyte differentiation antigen CD14 precursor	300	NM_000591
23	GAS2	Growth-arrest-specific protein 2	882	NM_005256
24	CASP8	caspase 8, apoptosis-related cysteine peptidase	650	NM_001228
25	CD38	Lymphocyte differentiation antigen CD38	393	NM_001775
26	AHR	Aryl hydrocarbon receptor precursor	2511	NM_001621
27	FAF1	FAS-associated factor 1	650	NM_007051
28	P53	Cellular tumor antigen p53 (Tumor suppressor p53)	393	NM_000546
29	FAS	Fatty acid synthase (TNFRSF6)	2511	NM_000043
30	PAX3	Paired box protein Pax-3	479	NM_000438

Approximately 54,000 new cases of non-Hodgkin's lymphoma are diagnosed annually in the US [Guo *et al,.* 2005]. Microarray technology has been used to improve both the accuracy of lymphoma classification and to identify oncogenic changes [Heisler *et al.,* 2005]. A CpG island (CGI) sequence library [Heisler *et al.,* 2005] containing 8,544 clones has been designed for studies of DNA methylation status (methylation cDNA microarrays). Methylation microarrays were used for detecting methylated genes in lymphoma [Guo *et al.,* 2005] and other types of cancers [Huang *et al.,* 1999]. Recent studies [Gopisetty *et al.,* 2006] have shown that DNA methylation is related to many cancers by silencing the genes involved in the apoptotic (cell death) pathway.

Cancer cells avoid apoptosis (death) by a number of mechanisms that include increased expression of anti-apoptosis proteins (cell do not die) and decreased expression (silencing) of tumor suppressor genes.

The set of 30 genes (see Table 3.7), henceforth denoted as GD30, are involved in cell death (apoptosis).

The genes in the first set (lines 1-10 in Table 3.7) are anti-apoptotic, that is, their expression corresponds to cell survival. The genes in the second set, (lines 11-20), are pro-apoptotic, that is, their expression signals cell death. The genes in the third set, (lines 21-30), are involved in apoptosis but their Gene Ontology annotations do not specify if they are anti- or pro-apoptotic. Although the above classification is somewhat simplistic, we assume it to be correct for the sake of illustrating the algorithms described in this book. For the GD30 genes we gathered the following data:

-gene name
-gene description
-gene amino acid (AA) sequence
-Gene Ontology (GO) terms. For each term we extracted:
   -term name
   -term GO code
   -term GO branch (molecular function-F, biological process-P, cellular component-C)
   -term evidence (Traceable Author Statement-TAS, Inferred from Direct Assay-IDA, etc.)

3.7.2 Gene clustering using Gene Ontology annotations

The right similarity measure for a problem depends both on the application and on the algorithm employed. Suppose we want to cluster the genes from the GD30 set. How many groups of genes can we identify in our dataset? To answer this question, we need to compute all pair-wise gene similarities (i.e. the gene similarity matrix) employing some GO similarity measure (such as FMS) and then use one of the clustering algorithm presented in Chapters 2 and 5. To make the dataset easier to analyze, we prearranged the GD30 set (first 10 are anti-apoptotic, next 10 pro-apoptotic and last 10 involved in apoptosis but not specified whether anti or pro) such that the three cluster structure reveals itself just by displaying the gene similarity matrix. This will allow us an initial analysis independent of any clustering technique. We would also like to compare the GO similarities introduced above with the traditional gene similarities techniques based on the amino-acid sequence. Can we see three clusters in the gene sequence similarity matrix (Figure 3.5), produced by the Smith-Waterman [Smith and Waterman 1981] dynamic programming algorithm? By inspecting the sequence similarity matrix from Figure 3.5 there does not seem to be **any** clusters in our data. Moreover, it seems that most of the genes from our example do not exhibit significant sequence similarity to any other gene in the data set.

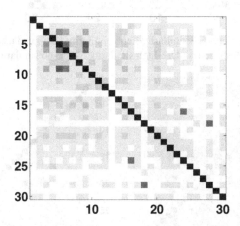

Figure 3.5 Sequence similarity (Smith-Waterman) between the 30 apoptosis genes (dark-high similarity, white-no similarity). As we can see from this picture, based on sequence similarity, the genes in our GD30 data set does not seem to be similar.

On the other side, by inspecting the GO term FMS similarity matrix (Figure 3.6) for the same set of genes we can clearly "see" two clusters: the first one (genes 1 to 10) that contains anti-apoptotic genes and the second one (genes 11 to 20) that contain pro-apoptotic genes. FMS is able to capture functional similarity between the genes in GD30 that could not be inferred based on the sequence similarity alone. We also note that neither the GO term normalized average similarity computed

using Equation 3.6 (Figure 3.7.a) nor the Jaccard similarity computed
using Equation 3.7 (Figure 3.7.b) reveal an obvious cluster structure.

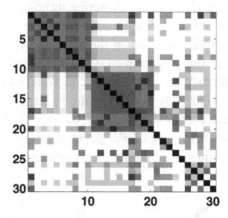

Figure 3.6 GO annotation FMS similarity (term similarity computed using Equation 3.3)
between the 30 apoptosis genes (dark-high similarity, white-no similarity).

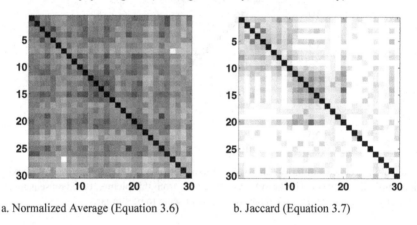

a. Normalized Average (Equation 3.6)   b. Jaccard (Equation 3.7)

Figure 3.7 GO annotation average normalized similarity and Jaccard similarity (term
similarity computed using Equation 3.3) between the 30 apoptosis genes (dark-high
similarity, white-no similarity).

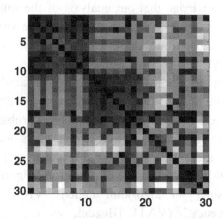

Figure 3.8 GO term Choquet similarity (Equation 3.15) for 30 apoptosis genes.

The Choquet similarity for the same set of genes (Figure 3.8) tends to reveal the cluster structure, too. However, it appears that this cluster structure is different than the FMS one, mainly in the third group of genes (index 20 to 30) where it exhibits a sub-cluster.

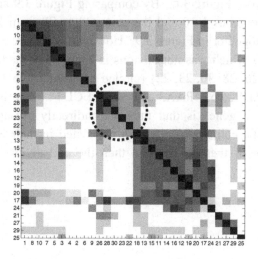

Figure 3.9 The image of the GD30 GO term FMS after applying VAT [Bezdek *et al.* 2002], (compare to Figure 3.6). The indices shown in this figure correspond to the ones from Figure 3.6 (that is, the gene on the second row with index 8 is the same as the one on row 8 in Figure 3.6).

So far, we acknowledge that our analysis of the similarity measures was somewhat speculative. Next, we will use the similarity matrices computed above to cluster our dataset employing some of the techniques introduced in Chapter 2.

Many clustering algorithms (such the C-means family) require, as an input, the desired number of clusters. However, usually the number of clusters in a data set is not known. To address this problem, we can run a clustering algorithm for a variety of cluster numbers and find the "best" one via a validity measure. We refer the reader to [Theodoridis and Koutroumbas, 1998] for a full range of cluster validity measures. In this chapter, we consider an algorithm called "Visual Assessment of (Clustering) Tendency" (VAT) [Bezdek *et al.*, 2002], which is particularly useful for inspecting the clustering structure of small data sets (hundreds of genes). The VAT is based on minimum spanning tree algorithm and outputs a permutation of the gene index that best reflects the cluster tendency of the data. As an example (Figure 3.9), we show the image obtained after VAT was applied to the FMS matrix of the GD30 genes from Figure 3.6. By comparing Figure 3.9 and Figure 3.6, we see that the genes 1-10 and the genes 11 to 20 from Figure 3.6 are still clustered together in Figure 3.9. However, in Figure 3.9 we observe an extra cluster which was not obvious in Figure 3.6, made of the genes with indices {26, 28, 30, 23, 22} (circled),. From Table 3.7 we find the genes to be: AHR, P53, PAX3, GAS2 and CD14. The interesting fact about the above genes is that they are indirectly pro-apoptotic (they contribute to the cell death) although not stated in their annotation set. That is, if one of them is mutated then the cell divides uncontrolled, leading to cancer.

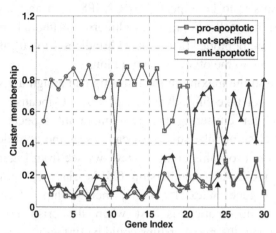

Figure 3.10 Cluster memberships computed using fuzzy C-means for the GD30 dataset using *C*=3 clusters: anti-apoptosis, pro-apoptosis and apoptosis, nos.

Once we have an idea of the number of clusters in our dataset, we apply a clustering algorithm, such as fuzzy C-means (see Chapter 2, Section 2.6.1), to determine the gene memberships in clusters. We mention that the traditional way of clustering objects described by a similarity matrix is to apply a relational clustering algorithm such as relational fuzzy C-means [Hathaway *et al.*, 1994] (see Section 5.2.3). However, in this case, we constructed feature vectors by using the similarity between genes as features [Claverie, 1999]; that is $G_i = (s(G_i, G_1), ..., s(G_i, G_{30}))^t$, $i=1...30$. Then, we applied the non-relational fuzzy C-means (see Chapter 5.2.3) to cluster the GD30 genes. The class memberships obtained on our GD30 apoptosis dataset are shown in Figure 3.10.

Using the cluster membership shown in Figure 3.10, we assign each gene to the cluster where its membership is maximum (that is, gene 1 will be labeled "anti-apoptotic", gene 11 - "apoptotic", etc.). From Figure 3.10 we see that the majority of the genes will be assigned as expected (that is, gene 1 to 10 in class one- "pro-apoptotic", genes 11 to 20 in class two - "anti-apoptotic", and genes 21 to 30 in class three- "apoptotic, not specified"). An interesting case is gene index 24 (CASP8) that has the maximum membership in the "pro-apoptotic" class and not in the "apoptosis, not specified" one as expected. Upon further investigation, this is to be expected as CASP8 is an apoptosis initiator gene - part of the cell apoptosis machinery. Although not directly annotated as "pro-apoptotic" (as genes with indices 1 to 10), its apoptotic role is implicit due to the other GO annotations.

If we only find the maximum membership and assign a given gene to a class, we do not take full advantage of fuzzy C-means. Employing fuzzy C-means for clustering has two important aspects. First, by inspecting the class memberships, we can assess the confidence of the class assignment. From this point of view, we see from Figure 3.10 that gene index 2 (ANXA1, membership 0.8 in class 1) is more "anti-apoptotic" than gene index 1 (RAF1, membership 0.5 in class 1). Obviously, this information is lost when the crisp assignment is performed. Second, the memberships could be further used in computing various class properties. An example of an application where we use fuzzy memberships is gene summarization [Popescu *et al.*, 2004].

3.7.3 Gene summarization using Gene Ontology terms

Gene summarization, or gene categorization [Joslyn *et al.*, 2004], consists of finding the main GO functions for a group of genes. In microarray experiments (see Chapter 5), for example, it is helpful to know the main functions of the over-expressed genes.

The fact that many gene products contain multiple functional domains suggests that fuzzy clustering may be a good methodology for summarization. When fuzzy clusters are found in similarity data representing sets of gene products, an individual gene product may have non-zero membership in more than one family, i.e., in multiple protein families. Hence, its annotation terms can contribute to the summary of all the clusters where it has a non-zero membership.

Let $U(C, N)$ be the $C \times N$ fuzzy membership matrix that results from clustering N gene products into C clusters using the fuzzy C-means, as explained in the previous section (visualized in Figure 3.10 for GD30). Let $P(N, NGO)$ be the $N \times NGO$ term membership matrix defined as:

$$P_{ij} = \begin{cases} IC(T_j) * c_{ij}(T_j, G_i) & \text{if GO term } T_j \text{ annotates gene } G_i \\ 0 & \text{else} \end{cases}, \quad (3.16)$$

where NGO denotes the number of distinct terms in the GO appearing in all gene products in the dataset, $IC(T_j)$ is the information content of term T_j, and $c_{ij}(T_j, G_i)$ is the reliability of the annotation of gene G_i with T_j. Then, we can summarize the function of N gene products by:

$$TM = U * P,$$ (3.17)

where TM is a $C \times NGO$ matrix with the ith row representing the contributions of all GO terms to the ith cluster. The representative term for the ith cluster is then defined as that term corresponding to the maximum value (weight) in the ith row, i.e. the term with the greatest contribution (in terms of IC and reliability) to cluster i. A final reliability value for a term in a cluster can be computed by normalizing the maximum value from row i using the sum of the row i.

For our apoptosis example $C=3$, $N=30$ and $NGO=139$. The summarization of the data set with three GO terms is shown in Table 3.8.

Table 3.8 Summarization of the apoptosis genes GD30 with three GO terms

Gene Index	{1...10}	{11...20}∪{24}	{21...23}∪{25...30}
Term ID	GO:006916	GO:006917	GO:006915
	(anti-apoptosis)	(pro-apoptosis)	(apoptosis, nos.)
Max weight	3.56	3.1	1.74
Reliability	0.14	0.12	0.05

As expected, the first cluster (genes 1 to 10) was summarized by the anti-apoptosis term (GO:006916), the second cluster (genes 11 to 20, and gene 24) was summarized by the pro-apoptosis term (GO:006917) and the last cluster (genes 21 to 30, except gene 24) was summarized by the generic apoptosis term (GO:006915). However, while their maximum weight (IC weighted by reliability, row 3 in Table 3.8) is somewhat high, their relative reliability (row 4 in Table 3.8) is low indicating that there might be some other candidates (not shown in Table 3.8).

3.8 Ontology Similarity in Data Mining

The ontological similarities described at the beginning of this chapter can be employed in a large variety of data mining algorithms instead of features and distances computed in some N-dimensional space. We will further refer to this class of algorithms as ontological algorithms. The only requirement for an algorithm to use ontological similarities is to be relational; that is, to employ only relative similarities (distances) between objects and not feature vectors that describe the objects in a feature space. This requirement is met, for example, by support vector machines (SVM) algorithms where the kernel is computed as the similarity between a set of objects. In fact, Ben-Hur and Noble [2005] used a GO-based kernel to find protein-protein interactions. Aside from the SVM, other examples of data mining algorithms where ontological similarities have been employed are: NERFCM (see section 5.2.3), Smith-Waterman [Gamalielsson and Olsson, 2005], and fuzzy rule systems [Popescu *et al.*, 2007].

As observed by Andreasen [Andreasen *et al.*, 2003], the similarity between two ontology terms can be interpreted as a fuzzy membership of one term in the concept denoted by the other term. This observation makes the case for considering every ontological algorithm as a fuzzy algorithm. Moreover, we believe that the fact that we literally compute the similarity using set of words and not feature vectors, makes every ontological algorithm a step closer to Zadeh's "computing with words" paradigm [Zadeh, 2002].

The clustering of gene products based on their GO similarity has been previously reported in bioinformatics [Popescu *et al.*, 2006; Pal *et al.*, 2005; Speer *et al.*, 2004]. We can use either hierarchical or NERFCM clustering for this purpose.

Gamalielsson and Olsson [2005] used a GO-based similarity between genes to align fragments from gene pathways with a modified version of the Smith-Waterman algorithm. For example, the pathway fragment composed of four gene symbols FAR1-SIC1-[   ]-CLN2-SIC1 can be aligned (with a gap) to the fragment FAR1-CLN1-SWI6-CLN2-SIC1 composed of five genes. Their approach (GOSAP) can be used to conduct searches in pathway databases and to map groups of microarray genes to pathways.

Popescu *et al.* [2007] employed gene similarity as fuzzy memberships to fire rules in an ontological fuzzy rule system (OFRS). Fuzzy rule systems (FRS) were described in detail in Chapter 2. Essentially, an OFRS is a Mamdani-Assilion FRS (MAFRS) where the membership values are computed using the semantic similarity between terms, instead of using standard membership functions. Popescu *et al.* used the OFRS to map genes to KEGG [Kanehisa and Goto, 2000] regulatory pathways. The fuzzy rule base consisted in rules of the form:

IF gene$_1$=G_1 AND...AND gene$_n$=G_n THEN pathway=P_k, (3.18)

where $\{G_i\}_{i=1,n}$ are the genes mentioned in KEGG as present in pathway P_k. In fact, the fuzzy rule base consists of the entire KEGG database for a given organism. To map a group of query genes $\{Q_j\}_{j=1,m}$ to a single pathway P_k that contains the genes $\{G_i\}_{i=1,n}$ (see Equation 3.18), one has to compute the GO-based similarity $s_k = sim(\{G_i\}_{i=1,n}, \{Q_j\}_{j=1,m})$ using, for example, Equation 3.6 twice. First, Equation 3.6 is used to calculate the pair-wise gene similarities based on their GO annotations, $s_{ij}(G_i, Q_j)$, as shown in Example 3.1. Here, we note that in calculating pair-wise term similarities we found that a path-based approach (Equation 3.1) was better than the one based on information content (Equation 3.4) [Popescu *et al.,* 2007]. Second, Equation 3.4 is applied to calculate s_k by aggregation of the pair-wise gene similarities $s_{ij}(G_i, Q_j)$.. Then, the pathway P_k will be inferred as likely for the input genes with a confidence (activation) s_k. However, there are typically hundreds of pathways (say M) for an organism in KEGG, each with an associated rule. Hence, we have to aggregate the output of all M rules to compute the OFRS output. As in the case of a MAFRS, a variety of aggregation strategies are possible. The simplest strategy is to choose the pathway P_m with maximum activation as:

$$m = \arg\max_{k=1,M}\{s_k\}.$$ (3.19)

However, since the KEGG pathways themselves form an ontology (as they are arranged in a hierarchy) we can further use the KEGG ontological similarity between pathways in the aggregation process. A possible aggregation strategy is to designate as output the nearest common ancestor of all the pathways activated over a certain limit. For example, if the P_1 = "Glycolysis metabolism" pathway was activated with $s_1 = 0.9$ and the P_2="Pyruvate metabolism" pathway was activated with $s_2 = 0.6$, then we can infer that the group of input genes belong to the "Carbohydrate metabolism" pathway, the NCA (P_1, P_2), with a likelihood $s = \min(s_1, s_2) = 0.6$.

Similar to the GOSAP algorithm, the OFRS may be used for mapping groups of genes to pathways or to conduct search through pathway databases. The GOSAP approach has the advantage of being more precise than the OFRS since it also accounts for the topology of the pathway. However, the tradeoff is between accuracy and computing speed, since GOSAP has to perform (expensive) comparisons to all the candidate subpathways while OFRS performs only one similarity assessment per pathway.

3.9. Discussion and Summary

In this chapter we discussed various aspects related to the link between fuzzy logic and ontologies. First, we showed that fuzzy logic can provide interesting new ways, such as the fuzzy measure similarity and the Choquet similarity, of comparing two objects annotated by an ontology. By expanding the set of ontological similarity tools available, we can choose the right tool for the job. For example, we mentioned that the widely used information content approach to computing term similarity was not found useful in applications where fuzzy inference was involved, such as ontological fuzzy rule systems.

Second, we demonstrated how ontological similarity together with existent fuzzy algorithms such as fuzzy C-means and NERFCM, can be used to solve bioinformatics problems. Several of the applications discussed here, such as gene clustering and gene summarization, are linked to microarray processing discussed in Chapter 5.

Lastly, we argued that the ontological similarity between two concepts may be interpreted as a fuzzy match between the same concepts. The acceptance of this interpretation opens a variety of possibilities in using fuzzy ontological matches in the computational intelligence algorithms. We showed an example of this approach, ontological fuzzy rule systems, where the rules are fired using ontological similarity instead of membership functions.

Chapter 4

Fuzzy Logic in Structural Bioinformatics

4.1 Introduction

This chapter focuses on the applications of fuzzy logic in characterization, comparison and prediction of protein structures. The basics of proteins and protein structures can be found in Appendix I. Each protein with a given amino acid sequence folds into a unique three-dimensional structure under physiological conditions [Anfinsen, 1973]. The three-dimensional structure of the protein, in turn holds the key in understanding the function of the protein at the molecular level. Traditionally, the structure of the protein is determined using experimental methods like X-ray crystallography and Nuclear Magnetic Resonance (NMR). Although these methods usually result in a high-resolution structures, they are both time consuming and expensive. It may take months or even years to determine the structure of one protein. Due to advancements in sequencing technologies, many complete genomes are being sequenced every year, producing millions of proteins whose structures need to be characterized. Obviously, the structures for most of these proteins will not be determined by experimental approaches.

An alternative to determining the structures experimentally is to predict the structure of a protein computationally. The input to the prediction system is the amino acid sequence (also called the primary structure) of the protein and the output is the three-dimensional structure of the protein. In spite of many advancements in the techniques used, these methods generally provide low-resolution structures. Nevertheless, low-resolution prediction results can provide useful insights for protein function and design. In addition, the prediction methods when combined with experimental methods may expedite the process of determining the structure dramatically while reducing cost. Since the three-dimensional structure of the protein is challenging to predict directly, researchers began to focus on the important intermediate steps such as secondary structure prediction and solvent accessibility prediction. Once the structure of a protein is predicted, one can characterize its function through structural features. Alternatively, the database of proteins with known structures and functions can be searched for a structural homolog (protein with high degree of structural similarity) to predict the function of the protein.

The three-dimensional structure of a protein can be predicted either directly from the primary sequence or using the secondary structure of the protein. Secondary structure plays an important role in characterizing protein structures and providing a basis for tertiary structure prediction [Rost, 2001; Meiler and Baker, 2003]. The secondary structure of the protein provides the computational methods with constraints that reduce the search space greatly and makes the prediction more efficient and faster. By predicting the secondary structure of the protein before predicting the tertiary, one mimics the natural order of events in the folding pathway, i.e., the secondary structure formation is often followed by folding the protein into a three-dimensional compact structure. Therefore, the study of secondary structure prediction is a crucial part in protein three-dimensional structure prediction.

Another important intermediate step that helps in characterization of protein tertiary structure is the prediction of solvent accessibility of the protein. The extent to which a solvent molecule can access the residue (an amino acid in the context of a protein) surface of a protein is called solvent accessibility. It sheds light on the packing of the residues, helping sequence alignments and tertiary structure prediction [Rost and Sander, 1994b; Rost et al., 1997].

The function of a protein can be inferred to a large extent, if it is similar to another protein whose structure and function are known. This inference can be achieved by comparing the structures of the proteins and obtaining a quantitative similarity. Therefore, efficient methods for protein structure comparisons are essential in bioinformatics. Protein structure alignment (also called protein structure comparison) can be defined as matching the three-dimensional geometry of protein backbones through a rigid-body transformation (rotation and translation). Secondary structures are usually weighted more during the matching process. Protein structure alignment can help in discovering the structure-function relationship in proteins. The structure similarity could shed light on the function of a novel protein that does not have any sequence homologs in the Protein Data Bank (PDB, [Berman et al., 2000]). Structure alignments also help in classifying proteins into families that may indicate the underlying evolutionary relationship among them. They can also help in evaluating the performance of protein tertiary structure prediction algorithms.

The knowledge of the structural class of a protein based on structural comparison provides insights into its function and its relationship with other proteins. One possibility is to classify a protein based on secondary and tertiary structures. SCOP (Structural Classification of Proteins) [Murzin et al., 1995], CATH (Class, Architecture, Topology, Homologous super family) [Orengo et al., 1997] and FSSP/DDD (Fold classification based on Structure-Structure alignment of Proteins/Dali Domain Directory) [Holm et al., 1992; Holm and Sander, 1996] are some examples of databases in this direction. The lowest level in SCOP is based on the arrangement of secondary structure elements and is called 'protein class' classification.

In this chapter, we emphasize the role of fuzzy logic in these important areas of structural bioinformatics by explaining one approach for each of the above mentioned areas. In addition to the applications explained in this chapter, other applications of fuzzy logic in structural bioinformatics have been proposed. For example, a fuzzy inference engine (see Chapter 2) was used in protein surface segmentation [Heiden and Brickmann, 1994]. Fuzzy cluster analysis (see Chapter 2) based on physicochemical properties of amino acids for secondary structure recognition was proposed by Mocz [1995].

4.2 Protein Secondary Structure Prediction

Each protein folds in to a compact three-dimensional structure, which is determined by its sequence. Each amino acid in this structure will adopt one of the following eight secondary structure classes: H (α-helix), G (3_{10}-helix), I (π-helix), B (isolated β-bridge), E (β-strand), S (bend), T (turn), and C (coil). Generally, researchers focus on a simplified version of the problem that contains only three secondary structure classes, such that {H, G, I}→H (Helix), {E, B}→E (Extended Strand), and {C, T, S}→C (Coil), according to the CASP standard (http://predictioncenter.org), where CASP is a community wide experiment on critical assessment of techniques for protein structure prediction. Given an amino acid sequence, the aim of protein secondary structure prediction is to computationally assign each residue one of the three secondary structure classes. An example is illustrated in Figure 4.1.

```
amino acid          ...QTLEGLFDDPNAETWAMKELLTGRLVFGENL...
secondary structure ...CCCCHHHCCCCCHHHHHHHHCCCCEEEECCC...
```

Figure 4.1 A protein fragment and its corresponding secondary structure.

Owing to the importance of protein secondary structure prediction, much attention has been given to this problem over the past three decades [Chou and Fasman, 1974; Qian and Sejnowski, 1988; Holley and Karplus, 1989; Zhang et al., 1992; Yi and Lander 1993; Rost and Sander, 1993; Rost and Sander, 1994a; Salamov and Solovyev, 1995; Chandonia and Karplus, 1995; King and Sternberg, 1996; King et al., 1997; Salamov and Solovyev, 1997; Rychlewski and Godzik, 1997; Karplus at al., 1998; Jones, 1999; Baldi et al., 1999; Ward et al., 2003; Jiang, 2003; Cheng et al., 2005; Bondugula et al., 2005]. Of all the successful prediction methods, the most popular systems are based on neural network techniques [Rost and Sander, 1994a; Jones, 1999], nearest neighbor (NN) approaches [Salamov and Solovyev, 1995; Salamov and Solovyev, 1997; Bondugula et al., 2005], hidden Markov model (HMM) methods [Karplus et al., 1998], and support vector machines [Ward et al., 2003; Cheng et al., 2005]. Among them, nearest neighbor methods are simple and transparent, and they do not require retraining when new data is available. Nearest neighbor methods are successful when sequences similar to the query sequence can be found in the PDB, but have limited performance otherwise. Although the nearest neighbor methods are sub-optimal techniques, the 1-NN rule is theoretically bounded above by no more than twice the optimal Bayes error rate [Cover and Hart 1967; Fukunaga and Hostetler 1975].

Many researchers have demonstrated [Rost and Sander, 1994a; Salamov and Solovyev, 1995; Jones, 1999] that secondary structure prediction accuracy could be increased by incorporating evolutionary information in the form of the Position Specific Scoring Matrix (PSSM) [Altschul et al., 1997]. PSSM is a profile of a protein that represents position-dependent amino acid distribution derived from the multiple sequence alignments. An amino acid at a particular position that is highly conserved receives a higher score than one that is less conserved at the same position. A protein of length l has a PSSM of dimension $l{\times}20$.

Neural network and HMM methods perform well if the query protein has many similar sequences in a sequence database to build a good PSSM but are less successful in other cases [Geourjon and Deleage, 1994; Salamov and Solovyev, 1995]. In addition, these methods may under-utilize the structure information in PDB when the query protein has some sequence similarity to a template (known structure) in the PDB, compared with the nearest neighbor methods. We will now describe a secondary structure prediction system that combines a generalized k-nearest neighbor (KNN) algorithm, the fuzzy k-nearest neighbor algorithm (FKNN) [Keller et al., 1985] and a neural network.

Hybrid models provide us with methods to combine the strengths of the individual methods and overcome their weaknesses to some extent. MUPRED [Bondugula and Xu, 2007] is a hybrid secondary structure prediction system that integrates the information from the FKNN algorithm and the PSSM using a neural network. The framework combines the strengths of the two methods: it uses the sequence profile information more effectively than template-based methods, while it also has a better potential to utilize the information in the PDB than PSSM-based methods. The system also provides a confidence measure for the predicted result, which enables users to identify regions of the protein for which the prediction is more likely to be accurate. The MUPRED web server can be accessed at http://digbio.missouri.edu/mupred.

MUPRED incorporates the PSSM of the query protein for secondary structure prediction through the PSI-BLAST [Altschul et al., 1997] program and the *nr* database (http://www.ncbi.nlm.nih.gov), which is a collection of non-redundant protein sequences. The calculated PSSM is used in generating two sets of features that are fed to the neural network. To generate the first set of features, the authors converted the PSSMs into vectors that are suitable for training neural networks. First, these values were scaled into the normalized profile in the range of [0 1] using the maximum and minimum in the PSSMs of all proteins in the database. Each position in the query sequence is represented by a 20-dimensional vector representing the likelihood of each amino acid occurring at that position. An additional bit is used to mark the termini of the protein, resulting in a 21-dimensional vector per position. These scaled PSSM values are converted into vectors suitable for neural networks using the sliding window scheme, i.e., the vector that represents the profile values of the current residue is flanked by its neighbors on the both sides. The rationale for this process is that the secondary structure of an amino acid is not only based on the current amino acid, but also on its neighbors. The number of residues that will be added on each side is determined by the window size W. The authors experimentally found that W= 13 worked the best. Therefore, the first feature set consists of 21x13 = 273 features per residue.

The second set of features is generated from the fuzzy k-nearest neighbor algorithm using the following procedure: the calculated PSSM is used to search for protein fragments in the local database that are similar to sub-sequences of the query protein using PSI-BLAST. Each fragment returned by the search is accompanied by several measures that include the percentage of identical amino acids between the sub-sequence of a query and the corresponding hit fragment, the number of positive amino acid substitutions, i.e., ones with similar biochemical properties (refer to Appendix I for details), the number of gaps inserted and the expectation value *Eval.* (see page 23). *Eval* measures the statistical significance that indicates the possibility that the current hit is obtained by chance in a particular database. The returned results obtained from the search are scored based on *Eval* as:

$$S = \max\{1,7 + \log_{10}(Eval)\}. \tag{4.1}$$

DEYRRLFEPFQLFEIPSYRSLL	DEYRRLFEPFQLFEIPSYRSLL		
...GRTWII.....SNPESKNRL	...CCEEEE.....EEEEEECCC	0.47	6.67
DEEKSKMLARLLKSSH......	CHHHHHHHHHHCCCC......	2.20	7.34
........PARITESEF....L	........CCCCCHHHH....H	2.20	7.34
.....VWVIAKSGISSQQQSMR	.....HHHHHHCCCCHHHHHHH	3.50	7.54
DDYQRTW...............	CCCCHHH...............	3.90	7.59
..KGRTWKPVILRINRAARCVR	..ECCEEEEECCCCCCCEEEEE	4.20	7.62
IENGR.................	EECCE.................	4.50	7.65
...............SERLRL	...............HHHHHC	4.80	7.68
DDHRTW...............	CCCCCC...............	5.00	7.70
(a)	(b)	(c)	(d)

Figure 4.2 Calculation of the membership value of each residue in secondary structure classes. The query protein is shown in the top row. (a) Database fragments from the PSI-BLAST matches; (b) corresponding secondary structures of the database matches; (c) corresponding expectation value of the hits (*Eval*); (d) distance scores of the hits calculated from their respective *Eval's* (Equation 4.1).

The above expression was designed so that it roughly emulates the notion of a 'distance'. Matching fragments whose similarities to the segments of query sequence are statistically significant have low expectation values and therefore low distances. Similarly, for matching fragments whose similarities are not significant, the distances are large. In the next step, these matches are labeled based on the classes to which the residues of the neighbors belong. Since the matches are obtained from the database that contains proteins whose experimental (PDB) structures are known, the labels are obtained from the database. If the residue of the neighbor that is aligned with the current residue is in a Helix state, the membership of the neighbor in the Helix class is '1' and '0' in Strand and Coil classes. These labeled neighbors are then used to calculate the membership value of the current residue in three classes.

These membership values represent the confidence with which the current residue belongs to the three secondary structure classes. Figure 4.2(a) illustrates the database fragments for a typical query protein. The highlighted column depicts the neighbors using the multiple sequence alignments of the hits with the query protein. The secondary structures, *Evals* and the distance scores corresponding to the database fragments are displayed in Figure 4.2(b) 4.2(c) and 4.2(d), respectively.

The secondary structure state of each residue can be predicted from class membership values of the neighbors with the FKNN algorithm. The following technique, adopted and modified from [Keller ct al., 1985], provides the procedure to calculate the membership values of the current residue from the labeled neighbors. Let $P = \{r_1, r_2, ..., r_l\}$ represent a protein with l residues. Each residue r has k-nearest neighbors, i.e., hit fragments that that have a residue aligned with the current residue (see Figure 4.2). Also, let u_{ij} be the membership in the ith class ($i \in \{Helix, Strand, Coil\}$) of the jth neighbor. For each r, the predicted membership value u_i in class i can be calculated using the following algorithm:

BEGIN
  Initialize $i = 1$.
  DO UNTIL (r assigned membership in all classes)
    Compute $u_i(r)$ using:

$$u_i(r) = \frac{\sum_{j=1}^{k} u_{ij}\left(1/S(r,r_j)^{2/m-1}\right)}{\sum_{j=1}^{k}\left(1/S(r,r_j)^{2/m-1}\right)}, \qquad (4.2)$$

  Increment i.
  END DO UNTIL
END

where $S(r,r_j)$ is the matching score of residue r with residue r_j.

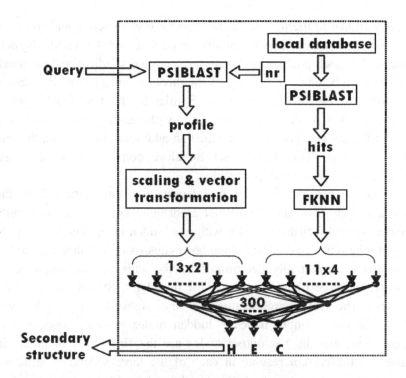

Figure 4.3 The block diagram of the MUPRED protein secondary structure prediction system. The profile of the query protein is used to generate two types of features. The first feature set consists of fuzzy class memberships of each residue in the three secondary structure classes. The second feature set consists of the normalized profile. The features are transformed into vectors suitable for neural network training using a sliding-window scheme of window length W. For the profile-derived feature-set, W=13 is used. An extra bit is used to mark the termini of each protein. The PSSM feature-set, therefore, consists of 13x21=273 features. For the fuzzy memberships, W=11 is used and, similar to the PSSM feature set, an extra bit is used to mark the termini of the protein, resulting in 11x4=44 features.

It can be noticed from Equation 4.2 (equivalent to Equation 2.14) that the contribution of each neighbor (hit r_j) in the calculation of membership value of the current residue in each class is determined by the score $S(r, r_j)$ (Equation 4.1), which in turn is determined by the significance of the hit in the PSI-BLAST search. The influence of the score can be controlled by the fuzzifier 'm' [Keller et al., 1985]. If the value of fuzzifier is set to 1.5, the class membership value of the residue is proportional to the inverse of the fourth power of score, and so on. In this case, the authors experimentally found that $m = 1.5$ yields the best results. For each position, there are three numbers indicating how much each residue belongs to the each of the three secondary structure classes according to the FKNN algorithm. Similar to the first feature set, a sliding window with W=11 was used by the authors to generate the second feature set. They also included an additional bit to mark the end of the protein. This feature set therefore consists of vectors that contain 11x4 = 44 features per residue.

A neural network is used to integrate the information from the normalized profiles and the FKNN algorithm. The network is a fully connected feed forward network with one hidden layer. The features are fed into the input layer. The hidden layer consists of 300 units, a number that was experimentally determined by the authors. The output layer consists of three nodes, one each for the Helix, Strand and the Coil classes. The final architecture of the network is as follows: $(273+44) \times 300 \times 3$ (input nodes × hidden nodes × output nodes). The values generated by the output nodes are the final class membership values of the current residue in each of the three secondary structure classes. The authors trained 100 networks and used the average value of the top four networks to determine the membership values. The procedure used in MUPRED is outlined in Figure 4.3.

MUPRED was trained and tested using a non-redundant set of proteins from the March 2006 release of PDBSelect [Hobohm and Sander, 1994] database. The PDBSelect database consists of proteins such that the sequence identity between any two proteins in the database does not exceed 25%. Initially the database had 3080 polypeptide chains. This database was filtered to select high-quality structures. In particular, only structures that are generated using the X-ray crystallography method with a resolution of 3 Å or less were selected. Of these, proteins with incomplete backbone atoms were discarded. Proteins that are shorter than 40 residues were also removed. Furthermore, if less than 90% of the protein residues are composed of regular amino acids, they are discarded too. Finally, the remaining 1998 proteins after the filtering constitute the local database of non-redundant structures. Of these 1998 proteins, the authors chose the oldest (according to the PDB release dates) 1000 proteins for tuning the parameters in the FKNN algorithm and for training the neural networks. The latest 200 proteins in this database were used as the first benchmark dataset (B1) to test and compare the performance of MUPRED with other secondary structure prediction systems. The training proteins contained 335,531 residues with 35.14% Helix residues, 23.75% Strand residues and 39.43% Coil residues. The authors used the Astral SCOP protein domain database version 1.69 [Brenner et al., 2000] to derive a second protein set for benchmarking purposes (B2). Each protein sequence of the original database, which contained 5457 protein domains, was searched for homologs in the training sets of MUPRED and other prediction software. If a homolog was found with a statistical significance value (*Eval*) of less than or equal to 0.1, the query sequence was discarded from the benchmark set. Similar to the earlier dataset, protein domain sequences that are shorter than 40 residues were removed and sequences that are composed of less than 90% of regular amino acids are discarded too. After this filtration process, only 1934 domain sequences remained in the second benchmarking protein set. The authors preferred the above method to evaluate and compare the performance of MUPRED with existing software to standard cross-validation schemes for the following reason: the earlier methods did not have access to large numbers of proteins, both for building the PSSM and the training data sets. The DSSP standard [Kabsch and Sander, 1983] of eight secondary structures were reduced to the CASP standard of three-state secondary structures as follows: {H, G, I}→Helix, {E, B}→Strand, and {C, T, S}→Coil.

There are three popular methods to measure the accuracy of secondary structure prediction systems. They are *Q*-measures [Rost and Sander, 1994a], Matthew's correlation co-efficient [Matthews, 1975] and Segment Overlap measure (SOV) [Zemla et al., 1999]. The authors used the first two measures to evaluate the performance of MUPRED and compared it with other existing software. The *Q*- measures are defined as follows:

$$Q_3 = 100 \times \frac{C}{T}, \tag{4.3}$$

$$Q_s = 100 \times \frac{C_s}{T_s},\qquad(4.4)$$

where C is the number of amino acids correctly classified in all three classes, T is the total number of amino acids, s is one of {Helix, Strand, Coil}. For example, C_{Helix} is the number of amino acids in Helix configuration that are correctly classified, while T_{Helix} is the total number of amino acids in the Helix configuration. Matthew's correlation coefficients are defined as follows:

$$M_s = \frac{TP \times TN - FP \times FN}{\sqrt{(TN + FN)(TN + FP)(TP + FN)(TP + FP)}},\qquad(4.5)$$

where s is one of {Helix, Strand, Coil}, TP is the number of positive cases that are correctly predicted, TN is the number of negatives that are correctly rejected, FP is the number of false positive cases, and FN is the number of false negative cases. For example, if a residue in a Helix is correctly predicted as Helix then it is a true positive case. If a non-Helix (either Strand or Coil) residue is correctly predicted as a non-Helix, then it is the case of true negative. If a Helix residue is predicted as a non-Helix residue, then it the case of false negative. Finally, if a non-Helix residue is predicted as Helix residue, it is case of false positive. Though most of the current secondary structure prediction methods produce a classification in terms in terms of fuzzy values (numbers in [0 1], representing the confidence in each of the three secondary structure classes), the metrics formulated for crisp classifications are used for accuracy assessment. Recently, measures for assessments of fuzzy predictions were introduced in [Lee, 2006]. The generalized forms of Q-measures are called F-scores, the generalized SOV is called fuzzy overlap measure (FOV) and the generalized M is called the fuzzy correlation coefficient. Detailed discussion on the formulae and applications of these generalized metrics can be found in [Lee 2006].

The authors compared the performance of MUPRED with PSIPREDv1[Jones 1999] and SSPro4[Baldi et al., 1999]. Both of them use PSSMs and neural networks and were also trained on the sets that contained similar number of sequences as in training set for MUPRED. The performance of the MUPRED version that just contained the FKNN algorithm (PSSM was used only to search the database) followed by a neural network filter was also reported. We present these results in Table 4.1. Except for the Q_{Helix} measure, MUPRED performed the best for all the other measures in both B1 and B2 datasets.

Table 4.1 The performance comparison of various algorithms on the two benchmark sets

Algorithm	Test set	Q_3	Q_{Helix}	Q_{Strand}	Q_{Coil}	M_{Helix}	M_{Strand}	M_{Coil}
FKNN+NN	B1	73.9%	76.2%	67.2%	76.1%	0.66	0.61	0.54
MUPRED	B1	**79.2%**	80.9%	**72.4%**	**82.0%**	**0.74**	**0.69**	**0.62**
PSIPREDv1	B1	75.9%	78.4%	68.3%	78.6%	0.70	0.63	0.56
SSPro4	B1	77.4%	**82.7%**	66.7%	79.5%	0.73	0.65	0.59
FKNN+NN	B2	76.1%	80.0%	68.2%	76.8%	0.69	0.63	0.57
MUPRED	B2	**80.1%**	83.9%	**72.6%**	**80.8%**	**0.75**	**0.69**	**0.63**
PSIPREDv1	B2	77.1%	80.2%	68.3%	79.0%	0.72	0.63	0.58
SSPro4	B2	78.4%	**84.4%**	67.3%	79.0%	0.74	0.65	0.60

Here, Q_3 is the fraction of amino acids whose secondary structures have been accurately predicted in all three classes. Q_{Helix}, Q_{Strand} and Q_{Coil} are the fraction of amino acids that are accurately predicted in Helix, Strand and Coil classes respectively. The numbers in bold indicate the best performance in the category. Similarly, M_{Helix}, M_{Strand} and M_{Coil} stand for Matthew's correlation coefficient for Helix, Strand and Coil classes respectively. B1- the 200 protein benchmark set derived from the March 2006 release of PDBSelect database. B2- the 1934 protein domain benchmark set derived from Astral SCOP database version 1.69.

The advantage of FKNN over the traditional (crisp) KNN algorithms is that residues are assigned a membership value in each class rather than binary decision of 'belongs to' or 'does not belong to'. Such an assignment allows the use of these membership values as (quantitative) strength or confidence with which the current residue belongs to a particular class. These strengths when fed to neural network along with the PSSM resulted in better performance when compared with existing methods. In the traditional (crisp) KNN, all the neighbors are weighted equally, which is not necessarily true in the context of proteins i.e., some protein fragments are more similar to the sub-sequences of the query protein than other fragments. This similarity is captured in the formulation of expectation value (*Eval*), which in turn is transformed into the score $S(r,r_j)$ in MUPRED. The FKNN was formulated such that these relative distances (score $S(r,r_j)$ in this case) are weighted when the query vector (current amino acid) is classified into one of the three secondary structure classes. The superiority of the FKNN over the traditional KNN algorithm for protein secondary structure prediction was also demonstrated in earlier work that lead to the development of MUPRED [Bondugula et al., 2005].

MUPRED is a simple and a novel framework that bridges the gap between the template based methods that find alignments between the whole query sequence or its short fragments and sequences in the protein structure database PDB and sequence profile based methods in which the sequence profile is derived from the similar sequences (typically without structural information). Template based methods are successful when sequences similar to the query sequence can be found in PDB, but have limited performance otherwise, mainly due to lack of using sequence profile information of the query protein. In contrast, sequence profile based methods take advantage of the sequence profile information but use the structure information in PDB indirectly. MUPRED overcomes this limitation by looking for fragments in the database that are similar to the segments of the query sequence rather than sequence-level homologs. Integrating these two fundamentally different models into a single model enables MUPRED to provide balanced predictions for queries with or without homologs in the sequence database. The notable feature of MUPRED prediction system is that the accuracy of the prediction increases as more and more protein structures become available without retraining or retuning. MUPRED also assigns confidences to the predictions, which enable the users to identify the regions of a protein for which the prediction is more likely to be accurate. The readers are referred to [Bondugula and Xu, 2007] for a complete description of the method to generate the confidence values and other details.

4.3 Protein Solvent Accessibility Prediction

The solvent accessibility of a protein may be defined as the extent to which the molecule of the solvent can access the residue in a protein. Generally, the residues of a protein are classified as either 'buried' (solvent molecules cannot easily access the residue; represented as B) or 'exposed' (solvent molecules can easily access the residue; represented as E). Sometimes, an additional class called 'intermediate' (I) is also used. The accessibility is often defined as the percentage of surface area for an amino acid that is exposed to the solvent. However, there are no universally accepted criteria to divide the residues into these classes. Predicting the solvent accessibility along with the secondary structure of a protein is an important intermediate step in the process of protein tertiary structure prediction. Solvent accessibility was demonstrated to assist in alignments in regions of low sequence identity for threading [Rost and Sander, 1994b; Rost et al., 1997]. Despite recent improvements, the prediction of solvent accessibility is less accurate when compared to the secondary structure prediction accuracy because the solvent accessibility is less conserved than the secondary structure.

In [Sim et al., 2005], a novel method for predicting solvent accessibility was proposed. The method uses profiles and a FKNN algorithm to predict solvent accessibility. The system was designed to make predictions in both two and three classes. For the two class classification into B/E, the following thresholds of solvent accessibility were used: 5% [i.e., (0, 5%) and (5%, 100%)], 16% and 25%. For three-class classification, the threshold of (0, 9%) for B, (9%, 36%) for I, and (36%, 100%) for E were used. The profiles of the proteins were generated using the PSI-BLAST program with the default parameters and the default scoring matrix. The low-complexity regions of the proteins, regions of a special structural conformation called coiled-coils, and transmembrane regions of the proteins were filtered out before running the prediction system.

For each residue, a sliding window of size 15 on the profile, centered on the current residue is used as the input feature vector to the FKNN algorithm. Sim et al. [2005] define a distance measure and weight to use in the FKNN algorithm. The distance between two feature vectors A and B is defined as:

$$D_{AB} = \sum_{i,j} W_i \left| P_{ij}^{(A)} - P_{ij}^{(B)} \right|, \qquad (4.6)$$

where $P_{ij}^{(A)} (i = 1,2,...,15; j = 1,2,...,20)$ is a component of the feature vector (note that P_{lj} represents the profile of the protein of l amino acids, while each j represents one amino acid), and W_i is weight parameter is defined as:

$$W_i = \left(8 - |8 - i|\right)^2. \qquad (4.7)$$

The weight is designed such that the current residue gets the maximum weight and importance of the neighbor i decreases with increasing distance from the current residue.

Sim et al. used a set of proteins derived from the Astral SCOP (version 1.63) chain-select-90 database as the reference. BLASTCLUST (www.ncbi.nlm.nih.gov/BLAST) was used to generate a training set. The authors eliminated any protein that was shorter than 50 amino acids in length. The resulting database contained 3460 non-redundant proteins with 819,090 feature vectors. Two additional datasets were used for benchmarking the performance. The RS126 [Rost and Sander, 1994a] database consists of 126 proteins. This served as the first benchmark dataset while another database that consisted of 229 newly added proteins to the Astral SCOP database was also included. Each database contains a non-redundant set of proteins such that the maximum sequence identity of any two proteins from any of the above three databases was at most 25%.

Sim et al. experimented with the various values for fuzzifier m and the number of neighbors to consider (k) in the FKNN algorithm (see Equation 4.2) and found the following optimal values: $(m, k) = (1.33, 65)$ for 3-state prediction (with 9% and 36% thresholds). For the two-state prediction (0%, 5%, 15% and 25% thresholds), the optimal values are $(m, k) = (1.5, 40), (1.25, 75), (1.29, 65)$ and $(1.33, 65)$, respectively. Two measures were used to assess the performance of the prediction system. The first measure indicates the percentage of correctly predicted solvent accessibilities, and the Matthew's correlation coefficient (Equation 4.5) was used as the second measure. The authors demonstrated that this method has a superior performance when compared to other methods that used neural networks, support vector machines and Bayesian statistics. For more details and performance issues, the reader is referred to [Sim et al., 2005].

4.4 Protein Structure Matching Using Fuzzy Alignments

Given two or more protein structures, the aim of an alignment algorithm is to return the new orientation of the aligned protein(s) with respect to a fixed reference protein by applying rigid body transformations on the protein being aligned. If $\mathbf{x_i}$ is the three-dimensional coordinates of the ith residue in the first chain and $\mathbf{y_i}$ is the three-dimensional coordinates of the jth residue in the second chain, the objective of the alignment methods is to minimize the sum of the distance $d(i,j)$ between the residue pairs on the two different chains. Usually, the alignment is performed using the C_α-atoms along the backbone using the following squared distance metric:

$$d_{i,j} = \left|\mathbf{x_i} - \mathbf{y_j}\right|^2 . \qquad (4.8)$$

As an example, the alignment between the inosine monophosphate dehydrogenase and glycolate oxidase is illustrated in Figure 4.4.

Most of the existing protein alignment methods use one of the two approaches: 1) directly minimize the inter-atomic distance between the aligned structure backbones (global alignment) or 2) minimize the distance between the structural segments of the proteins that need to be aligned (local alignment). We will now describe a method that uses the former approach, aided by a fuzzy weight matrix.

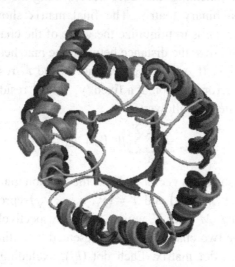

Figure 4.4 Structure alignment between glycolate oxidase (1gox, in green and red) and inosine monophosphate dehydrogenase (1ak5, in blue).

To align the protein structures, Blankenbecler et al. [2003] used a method based on the Needleman-Wunsch global sequence alignment algorithm [Needleman and Wunsch, 1970]. They modified the original algorithm to generate a fuzzy alignment matrix instead of a binary matrix. The advantage of using a fuzzy alignment matrix is that results allow for a probabilistic interpretation without using complex simulation techniques. The approach is also less sensitive to the choice of the distance measure as the distances are weighed by the fuzzy alignment matrix. Finally, user specified constraints may be easily incorporated in to the fuzzy alignment matrix.

The approach by Blankenbecler et al. [2003] is mainly an iterative two-step procedure. In the first step, a fuzzy assignment matrix W is calculated. Each element $W_{ij} \in [0,1]$ indicates the confidence that that the atom i in the first chain is matched with the atom j in the second chain. In the second step, one of the chains is translated and rotated with rigid body transformations using the W calculated in the first step. At the start of the algorithm, the degree of fuzziness is high and as the algorithm proceeds through the number of iterations, the degree of fuzziness is reduced through annealing, therefore transforming the fuzzy assignment matrix W into a binary matrix. The final matrix shows the matched atoms. The objective is to minimize the value of the chain error function E_{chain} i.e., to minimize the distance between the matched elements of the two protein chains. If x_i is the coordinate of the ith residue in the first chain, the chain error function at a fixed y_j (the jth residue in the second chain) is formulated as follows:

$$E_{chain} = \sum_{i=1}^{M} \sum_{j=1}^{N} W_{ij} \left(a + Rx_i - y_i\right)^2, \qquad (4.9)$$

where a is the translation vector and R is the rotation matrix.

Let $X = (x_1, x_2, ..., x_M)$ and $Y = (y_1, y_2, ..., y_N)$ represent two protein chains containing M and N amino acids, respectively. All possible alignments of the two chains can be represented by a directed path on an $M \times N$ alignment dot matrix. Each dot (i,j), excluding the dots at the boundaries, have $k=3$ possible predecessors along the alignment path. If $k=1$, a gap in chain X was aligned with an atom in chain Y. If $k=2$, an atom in chain X was aligned with an atom in chain Y, and if $k=3$, a gap in chain Y was aligned with an atom in chain X. The alignment cost D_{ij} is given by:

$$D_{ij} = \min_k \{\tilde{D}_{i,j;k}\},\qquad(4.10)$$

where $\tilde{D}_{i,j;k}$ is the alignment cost if the alignment of the path passes through the node given by k and can be calculated using the following recursive relations:

$$\tilde{D}_{i,j;k=1} = \min_{i \le l \le j}\{D_{i,j-1} + \alpha(l)\},$$

$$\tilde{D}_{i,j;k=2} = D_{i-1,j-1} + d_{i,j},\qquad(4.11)$$

$$\tilde{D}_{i,j;k=3} = \min_{i \le l \le i}\{D_{i-1,j} + \alpha(l)\},$$

where $\alpha(l)$ is the gap penalty for a gap of length l and $d_{i,j}$, as defined in Equation 4.8. Notice that $D_{M,N}$ holds the information only to compute the cost of the alignment of the two chains, the optimal path(s) are stored in another matrix $v_{i,j;k}$. Each element $(i,j;k)$ indicates the probability that the optimal path passes through (i,j) and through the preceding node based on k. The matrix v can be calculated using the following expression:

$$v_{i,j;k} = \frac{e^{-\tilde{D}_{i,j;k}/T}}{\sum_{k'} e^{-\tilde{D}_{i,j;k'}/T}},\qquad(4.12)$$

where $T>0$ is a parameter to control the fuzziness of the alignment matrix. For a large T, all paths are equally probable, indicating maximum fuzziness, while the limit $T \to 0$ results in one optimal path (the original Needleman-Wunsch algorithm). As the algorithm proceeds in several iterations, the parameter T is annealed and gradually the fuzzy matrix v will becomes a binary matrix indicating only one alignment.

    Blankenbecler et al. [2003] limited their method to the position-dependent linear gap penalties of the form:

$$\lambda_a^{(n)} + (l-1)\lambda_{ext},\qquad(4.13)$$

where, $\lambda_a^{(n)}$ is the gap opening penalty in protein chain n and λ_{ext} is the gap extension penalty for a gap of length l. The equation for calculating the alignment cost can be redefined in terms of v, $\lambda_a^{(n)}$ and λ_{ext} as:

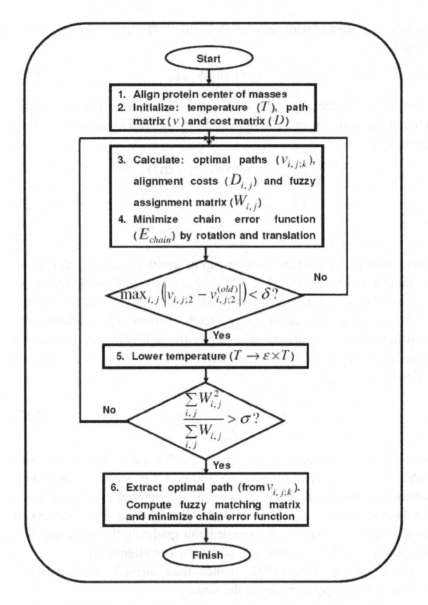

Figure 4.5 The flowchart of the protein structure matching using fuzzy alignments.

$$\tilde{D}_{i,j;1} = D_{i,j-1} + \lambda_j^{(2)}\left(1 - v_{i,j-1;1}\right) + \lambda_{ext} v_{i,j-1;1},$$

$$\tilde{D}_{i,j;2} = D_{i-1,j-1} + d_{i,j},$$

$$\tilde{D}_{i,j;3} = D_{i-1,j} + \lambda_i^{(1)}\left(1 - v_{i-1,j;3}\right) + \lambda_{ext} v_{i-1,j;3}, \tag{4.14}$$

resulting in an alignment cost at node (i,j),

$$D_{i,j} = \sum_k v_{i,j;k} \tilde{D}_{i,j;k}. \tag{4.15}$$

The probability that a node (i,j) is a part of the optimal path can be calculated using a recursive definition. With an initial value of $P_{M,N} = 1$, we get:

$$P_{i,j} = v_{i,j+1;1} P_{i,j+1} + v_{i+1,j+1;2} P_{i+1,j+1} + v_{i+1,j;3} P_{i+1,j}, \tag{4.16}$$

with the necessary condition $P_{0,0} = 1$. Using the definition of $P_{i,j}$ and $v_{i,j;2}$, the fuzzy assignment matrix can be defined as:

$$W_{i,j} = P_{i,j} v_{i,j;2}. \tag{4.17}$$

The above expression can be interpreted as the product of the confidence that (i,j) is part of the optimal path and the confidence that this pair is locally matched. The flowchart depicting the above process is illustrated in Figure 4.5.

The performance of this method was assessed using a set that covered a wide range of protein families and includes matching proteins with insignificant sequence similarity. Most of the protein pairs belonged to the same SCOP (http://scop.mrc-lmb.cam.ac.uk/scop) superfamily. Blankenbecler et al. chose pairs that had diverse structures and reported difficulty of alignment. Most of the protein pairs used to assess the performance were in the benchmarks of previous protein alignment studies. The authors compared the results of their method with following three popular protein alignment servers: Yale Alignment Server ([Gerstein and Levitt, 1996], http://molmovdb.mbb.yale.edu/align), Dali ([Holm and Sander, 1996], http://www.ebi.ac.uk/dali, and CE ([Shindyalov and Bourne, 1998], http://cl.sdsc.edu). The method employs two sets of parameters. The first set is used to fine tune the algorithm itself and the second set is used to explore the optimal gap penalties for alignments. The parameters for fine tuning the algorithm were derived from the ten training sequences that were randomly chosen from the benchmark set described above. The gap penalties were optimized to obtain the alignments in which the number of aligned atoms was comparable to that of the other methods. Once these parameters were set, they were left unchanged throughout the study. The performance of the method was measured in two terms: RMSD (root of the mean squared distance) between the aligned proteins and N, the number of atoms aligned in the proteins. In most cases, this approach produced superior performance when compared to the other techniques. It allowed for a fast alignment of proteins with the execution time scaling with the square of the length of proteins chains aligned. The reader is referred to [Blankenbecler et al 2003] for algorithmic and performance details.

4.5 Protein Similarity Calculation Using Fuzzy Contact Maps

Another approach for comparing protein tertiary structures is through mapping patterns of interactions among the residues. Two residues are said to be in contact if they are separated by a distance less than a predefined threshold. A matrix that records these contacts for all possible pairs of residues in a given protein sequence is called a contact map. A contact map provides invaluable information about the non-local contacts that help proteins form and maintain stable structures. While a contact map does not contain all information about the protein, it can be viewed as a good two-dimensional representation of protein three-dimensional structure. Contact maps are useful in protein three-dimensional prediction and for protein structure comparison [Carr et al., 2002; Caprara et al., 2004]. We will now briefly describe a work by Pelta et al., [2005] that introduces fuzzy contact maps and their application to protein similarity calculation.

A contact map for a protein of length l is a binary matrix of dimension $l \times l$ such that each element (i,j) is equal to 1 if the Euclidean distance between residues i and j is less than a pre-defined threshold T, or equal to 0, otherwise. The similarity between two proteins can be calculated by aligning their contact maps. Given contact maps of two proteins, the problem of finding their largest common sub-structure is called the 'Max-CMO' (Maximum Contact Map Overlap) problem.

These contact maps with a fixed threshold have some limitations. They cannot handle the uncertainties inherent in the determination of the atomic Cartesian coordinates by X-Ray crystallography or NMR. Some experimental errors can range from 0.01 to 1.27 Å which is close to some covalent bonds [Laskowski, 2003]. A contact map calculated at a fixed threshold loses information of contacts at other thresholds, i.e., it cannot characterize the distance of a contact between two residues well. To alleviate the above mentioned problems, Pelta et al., [2005] introduced the concept of fuzzy contact maps. In fuzzy contact maps, both the threshold and the meaning of the term 'contact' are generalized. First, the traditional contact maps are generalized by removing the constraint of having a single-threshold. This is useful in distinguishing various features of the protein. For example, the α-helices and β-strands have quite different contact patterns and can be conveniently represented using different thresholds. The second generalization is for the term 'contact' itself. In a traditional sense, a contact means anything that is less than threshold T. Fuzzy contact maps facilitate alternate definitions of contact like 'slightly more or slightly less than T' and 'T and slightly more than T'. The flexibility of the fuzzy contact maps helps the user to decide how much biology to include in these mathematical constructs.

A generalized maximum fuzzy contact map overlap problem or GMax-FCMO takes multiple thresholds and multiple definitions of the contact into account, while calculating the overlap of contact maps for aligning two protein structures. In GMax-FCMO, the objective is to maximize the number of alignments of pairs of residues that are in contact with respect to the same threshold and same semantic meaning of contact. The authors proposed the use of FANS, a fuzzy sets-based extension to the classical variable neighborhood search (VNS) [Hansen and Mladenovic, 2001] to solve the GMax-FCMO. In VNS, the neighborhood around the current solution, used to sample the solution space, is systematically and dynamically adjusted to allow the local search to proceed beyond a local optimum. In FANS, a fuzzy objective function is used to evaluate and accept moves in solution space. Pelta and his colleagues first calculate an initial alignment of the fuzzy contact maps and improve the solution by randomly applying one of the three neighbor operators (N1, N2 and N3), defined as follows:

- N1: inserts one random alignment into current overlap
- N2: inserts two random alignments into current overlap
- N3: changes the alignments in the current overlap to either left or right.

One neighborhood operator is randomly selected (with same probability for all operators) and applied to the current alignment. This process is repeated k times (initially, k=3). If all the neighboring solutions (new overlaps generated by applying neighborhood operators to current overlap) are worse then the current solution (based on the user defined fuzzy objective function), the value of k is reduced by 1. Essentially, this procedure first explores for good solutions in the neighborhood and if better solutions are not found, the current best solution is exploited further. The objective function used by the authors also allowed them to accept solutions that are sometimes worse then the current solution. The authors demonstrated that the solution using GMax-FCMO produced results that are comparable and sometimes superior, when judged against existing methods. Further details of the algorithms and discussion of FANS can be found in [Pelta et al., 2005].

4.6 Protein Structure Class Classification

One of classification criteria for a globular protein (a protein that is soluble in water) is based on its secondary structure composition. The four main categories are all-α, all-β, α+β and α/β. A protein is classified [Chou, 1989] as an all-α protein if it contains >45% α-helices and <5% β-strands, an all-β protein if contains <5% α-helices and >45% β-strands, an α+β protein if it contains >30% α-helices and >20% β-strands with dominantly anti-parallel β-strands (see Appendix AI.2), and an α/β protein if it contains >30% α-helices and >20% β-strands with dominantly parallel β-strands. The information about the structural class of proteins has a proven impact on protein secondary and tertiary structure predictions [Chou, 1989; Deleage and Roux, 1989; Cohen and Kuntz, 1987; Carlacci et al., 1991]. The class of a protein can be determined using its amino acid composition, hydrophobicity pattern of the residues or α-helix/β-strand content. We will now describe a method described in [Zhang et al., 1995] that uses amino acid composition and fuzzy c-means (FCM) clustering [Bezdek, 1981] (see Section 2.6) for protein class prediction.

Let S be a set that represents all globular proteins. Here, S contains four subsets S_α, S_β, $S_{\alpha+\beta}$ and $S_{\alpha/\beta}$, representing each of the four structure classes. If traditional (crisp) sets are used to represent these classes, then each protein belongs to only one subset. However, some proteins may possess characteristics of more than one subset, making traditional crisp sets inadequate to represent such natural phenomenon. Using fuzzy sets to represent structural classes is one possible solution to overcome the limitation imposed by crisp sets. A protein can belong to more than one structural class by having non-zero membership value in more than one class. Given such a flexible representation, two questions remain: 1) how can the memberships of a given protein in various structural classes represented by the fuzzy sets be determined and 2) given the membership value of a particular protein in various fuzzy sets, how can a specific structural class be assigned? The first question can be addressed by using the FCM to calculate the membership values of each protein in the four structural classes. There are many approaches to address the second question, and the simplest of all is to assign a protein to class in which it has the maximum membership value.

Each protein is represented by a twenty-dimensional vector f, such that each component of the vector represents the contribution of a specific amino acid to the composition of the protein. The vector f is defined as:

$$f_j = \frac{n_j}{\sum\limits_{i=1}^{20} n_i} \quad (j = 1,2,...,20), \qquad (4.18)$$

where n_j is the frequency of occurrence of the jth amino acid in the protein. Similarly, a set of i proteins x_i can be represented by:

$$F(x_i) = [f_1(x_i), f_2(x_i),..., f_{20}(x_i)], \qquad (4.19)$$

while the class prototype can be represented by:

$$F(k) = [f_1(k), f_2(k),..., f_{20}(k)], \qquad (4.20)$$

where $k = \alpha, \beta, \alpha+\beta$ and α/β. One of the easiest ways to determine the initial values of prototypes before running the FCM is to set each of them to the average value of the members within the structural class of the training data. Zhang et al. [1995] chose this approach here. The degree of membership of each protein in the four structural classes can be represented as $u_\alpha, u_\beta, u_{\alpha+\beta}$ and $u_{\alpha/\beta}$. While running the FCM, the following objective function was minimized over a set of n proteins with respect to the fuzzy membership values $u_k(x_i)$ and the 'fuzziness' index q (similar to Equation 2.7):

$$J_m = \sum_{k=1}^{4} \sum_{i=1}^{n} [u_k(x_i)]^q d^2[F(x_i), V_k], \qquad (4.21)$$

where V_k represents the cluster centroids that were defined as:

$$V(k) = [v_1(k), v_2(k), ..., v_{20}(k)], \qquad (4.22)$$

with $k = \alpha, \beta, \alpha+\beta$ and α/β. In addition, the solution must satisfy the following constraints:

$$0 \le u_k(x_i) \le 1, \qquad (4.23)$$

and

$$u_\alpha(x_i) + u_\beta(x_i) + u_{\alpha+\beta}(x_i) + u_{\alpha/\beta}(x_i) = 1. \qquad (4.24)$$

The Minkowski distance is used to calculate the distance between a given protein $F(x_i)$ and the cluster centroid V_k. The metric was defined as:

$$d[F(x_i), V_k] = \left\{ \sum_{l=1}^{20} |f_l(x_i) - v_l(k)|^p \right\}^{1/p} \qquad (4.25)$$

where p determines the particular Minkowski metric used: $p=1$ represents the 'city block' distance while $p=2$ represents the 'Euclidean' distance. Following the algorithm from [Bezdek, 1981], Zhang et al. used the FCM algorithm (see Section 2.6) and calculated J_m based on Equation 2.7 for clustering the proteins into the four structural classes.

Different choices for the initial cluster centers often lead to different clusters when using the FCM. This problem was eliminated by Zhang et al. by setting the initial representatives of the clusters to the average value of the members within the structural class of the training data that consists of 64 proteins used in [Chou, 1989]. Once the membership values of all the proteins in each of the four structural classes are calculated, the proteins are assigned to the class in which they have the maximum membership value. The accuracy of the algorithm was measured in terms of the percentage of correct predictions (PCP) of all the predictions. The parameters p (for use in Minkowski metric) and q (fuzziness index) were optimized to perform well on the 64 training proteins using the following random sampling method: the FCM was run several times using a combination of a random value for m in [1.0 3.0] and random value for m in [1.0 2.0] and the PCP was calculated. The combination that performed well on that training set was: $(p, m)=(2.4,1.4)$.

Zhang et al. also introduced a representation of the proteins in the membership-function space. The membership function space with four classes can be represented by a regular tetrahedron of height h, such that each of the four classes will be represented by a vertex. If the proteins are assigned crisp membership values, they can be visualized to be present at the vertices only. If the flexible fuzzy representation is used, the proteins can be present anywhere in the tetrahedron. The centroid of the tetrahedron corresponds to the state of maximum fuzziness. Representation of proteins in membership space serves an excellent aid in visualizing the clusters of the given proteins. The Cartesian coordinates of a particular protein in the tetrahedron of height h can be calculated from its membership values in the four structural classes using the following equations:

$$X = \frac{1}{2h}\left\{2\left[m_\alpha(x)+m_{\alpha+\beta}(x)\right]-1\right\}$$

$$Y = \frac{1}{2h}\left\{2\left[m_\alpha(x)+m_\beta(x)\right]-1\right\} \qquad (4.28)$$

$$Z = \frac{1}{2h}\left\{2\left[m_\alpha(x)+m_{\alpha/\beta}(x)\right]-1\right\}$$

In addition to achieving accuracy that was comparable with other methods on the training proteins, the authors reported that the method also performed well on blind test sets. The reader is referred to [Zhang at al., 1995] for further details.

4.7 Summary

This chapter discussed the applications of fuzzy concepts and methods in characterization and prediction of various protein structure features, including predicting secondary structure, solvent accessibility, contact map, and protein class, as well as comparing a pair of protein structures. Protein structures do not have simple geometry and typically contain some flexibility, which are suitable for fuzzy approaches to describe. The examples illustrated in this Chapter have demonstrated that fuzzy approaches are especially powerful for bioinformatics related to protein structures.

Chapter 5

Application of Fuzzy Logic in Microarray Data Analyses

5.1 Introduction

A living organism often has trillions of cells, each carrying the same genome. However, only a fraction of the genes coded by the genome are active in any given cell. These genes are "expressed" for the function of a cell. "Gene expression" is typically referred to as the transcription (see Appendix 1) of mRNA. Gene expression level changes over time and in response to environmental stimuli. For example, a bacterium often expresses more genes that help digest nutrients when it is in a nutrient-rich environment. Gene expression patterns may also change dramatically at different stages in the life cycle of an organism. A butterfly comes from a caterpillar, which has the same genome. However, they look very different and this is mostly due to different gene expression patterns. By knowing the expression levels of mRNA under different conditions and over time, one can infer extensive information about gene functions, gene regulations, and gene interactions.

DNA microarrays (a.k.a. biochips, DNA chips, gene arrays, genome chips) are currently the most popular technology to measure gene expression level. Microarray technology is based on the principle of DNA hybridization, that is, two single stranded DNA fragments tend to bind together (renaturation) at regions with sequence complementarity (i.e., A-T and G-C for DNA; A-U and G-C for RNA). A microarray often has tens of thousands of spots, each containing complementary

DNA molecules of the whole or part of a gene sequence as a probe. The probes are deposited on a solid surface such as glass or plastic. Microarrays utilize the preferential binding of complementary nucleic acid sequences between a probe and the gene that the probe represents. When the targets (DNA fragments from the unknown sample, fluorescently labeled) are deposited onto the array, they will "stick" (hybridize) only to the complementary probes. As a result of the hybridization process, the intensity of the fluorescent emission at a given probe location is directly proportional to the amount of DNA fragments with the same identity existent in the sample. In this fashion, a microarray can simultaneously monitor expression levels for most of the genes in a genome in a single experiment (see Figure 5.1).

Figure 5.1 A segment of microarray image (figure from http://doegenomestolife.org by the U.S. Department of Energy Genomics:GTL Program). Each spot represents a gene and the intensity represents its expression level.

There is a great variety of microarrays depending on their chip technologies, their probe types and their experimental designs. For instance, DNA probes are used in genomic applications while protein probes are used in proteomics applications, typically for inferring protein-protein interactions. In this section we describe two main variants of the microarray technology: cDNA microarrays and oligonucleotide (oligo) microarrays. In a cDNA microarray, the probes (cDNA fragments 500 to 5000 bases long) are extracted from a cDNA library and deposited on the chip support (glass) using a robotic device. There are usually two types of targets (mixed in a "soup") hybridized (washed over) at any given location in a cDNA array, one coming from "abnormal" cells and the other extracted from "normal" cells. To differentiate between them, the two target types are labeled with different fluorescent dyes; for instance, red for "abnormal" and green for "normal". As a consequence, if only green light is observed at one location, the DNA fragment (i.e. the gene) represented by the probe from that location is absent from the "abnormal" cell. This could be the case of a tumor suppressor gene that is missing in a cancerous cell. Similarly, if only red light is observed at a location, the given gene is missing in the "normal" cell. This could be the case of an oncogene (tumor causing gene) present in the cancer cell. The advantage of cDNA microarrays is that they can be customized to a given experiment. The disadvantage is that their quality highly dependent on the conditions existent in each lab.

In the oligonucleotide (oligo) microarrays, the probes are DNA fragments 20-80 bases long synthesized vertically on the chip using a technology (photolithography) inspired from the electronic component industry. The advantage of this type of array is that it tends to have a higher quality than a cDNA microarray. However, unlike the cDNA arrays they are not easily customizable. Other than these two types of microarrays discussed above, there are ChIP-chip arrays, tiling arrays, Affymetrix arrays, etc.

Microarrays have been used in various biomedical applications such as gene discovery [Chu *et al.*, 2005], disease diagnosis [Gabriele *et al.*, 2006], phamacogenomics (drug discovery) [Levy, 2003], and toxicology [Vrana *et al.*, 2003]. For example, microarrays are used to identify disease genes by comparing gene expression patterns in disease and normal cells. They can also be used as a diagnosis tool by checking possible abnormal gene expression for a disease. Readers can find more information about microarrays in [Baldi and Hatfield, 2002; Blalock, 2003; http://www.ncbi.nlm.nih.gov/About/primer/microarrays.html].

In this chapter we review several fuzzy processing algorithms for microarray data such as fuzzy C-means, relational fuzzy C-means and fuzzy co-clustering for gene selection and patient classification.

5.1.1 Microarray data description

The typical microarray dataset is shown in Table 5.1. It contains M samples (microarray chips, patients), each sampling the expression of N (typically from thousands to tens of thousands) genes. The expression values of a gene across different samples (a row in Table 5.1) are called the "gene expression profile". The expression profile of gene i is a vector $\mathbf{x}_i \in R^M$.

Table 5.1 Gene expression data matrix.

	Sample 1	...	Sample j	...	Sample M	Gene Expression Profile
Gene 1	x_{11}	...	x_{1j}	...	x_{1M}	$\mathbf{x}_1$
...	...	...	...	...	...	...
Gene i	x_{i1}	...	x_{ij}	...	x_{iM}	$\mathbf{x}_i$
...	...	...	...	...	...	...
Gene N	x_{N1}	...	x_{Nj}	...	x_{NM}	$\mathbf{x}_N$
Patient Expression Profile	$\mathbf{p}_1$	...	$\mathbf{P}_j$	...	$\mathbf{p}_M$	

Each microarray chip samples the expression of N genes. The dataset contains M ($\ll N$) samples (microarray chips). Row i is the expression profile of gene i and is represented by a vector $\mathbf{x}_i \in R^M$; column j is the expression profile of sample (patient) j and is represented by a vector $\mathbf{p}_j \in R^N$.

A typical expression profile is represented in Figure 5.2. Similarly, we can define a sample (patient) expression profile (a column in Table 5.1). The expression profile of patient j is a vector $\mathbf{p}_j \in R^N$.

To obtain a microarray expression dataset similar to the one shown in Table 5.1, several preprocessing steps, such as image processing and data normalization, are required. The image processing and the DNA fragment count for each spot is usually performed by the software included in the microarray reading device. The microarray normalization (both across genes and across samples) depends on the experimental design method and it is usually performed on a case-by-case basis. The normalization step deserves a special consideration, particularly for the case of cDNA microarrays where there is more flexibility in the experimental design methodology. In this chapter, we assume that the dataset has been normalized and has the form shown in Table 5.1. For more details about microarray experimental design and normalization, we refer the reader to [Quackenbush *et al.*, 2002; Yang *et al.*, 2002; Lee, 2004]. For a cDNA microarray the normalization step includes the merging of the two channels labeled with different dyes (red and green). A popular merging strategy is to compute at each spot the value $x_{ij} = \log_2(x_{ij}^{red} / x_{ij}^{green})$ where x_{ij}^{red} and x_{ij}^{green} is the median intensity of the related channels at location *ij*.

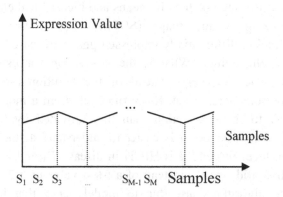

Figure 5.2 Typical gene expression profile. A gene can be represented as a 2-D curve or as a point in R^M.

The samples that represent the same condition of the organism (say follicular lymphoma), but extracted from different individuals, are called biological replicates. Often, the samples from a dataset represent several conditions (typically two; for example, "follicular lymphoma" and "normal" patients), each containing a number of biological replicates. An alternative way of obtaining a dataset similar to the one shown in Table 5.1 is to sample the same cell population at different points in time. Consequently, a time series of expression values (usually 10 to 20) for each of the *N* genes is obtained.

In the majority of applications, microarrays are used to address the following issues: (1) find the differentially expressed genes between conditions, as found in gene selection or biomarker identification, (2) find natural groupings among genes, conditions or both, (3) train a classifier to recognize a given condition using gene expression, and (4) find the regulatory relationship among a given set of genes.

5.1.2 Microarray processing algorithms for gene selection and patient classification

Traditional methods for finding genes that have a high expression in one condition and a low expression in another (referred to as differentially expressed genes) are statistical tests (such as t-test and Wilcoxon rank sum test), crisp clustering (such as K-means and hierarchical clustering), and self organizing feature maps (SOFM) [Lee, 2004]. Fuzzy approaches to finding differentially expressed genes are based on fuzzy rules and fuzzy clustering. What is the motivation for using fuzzy approaches for gene discovery instead of the traditional ones? We mention three reasons here. First, fuzzy rules represent a more human-readable method for gene selection than that provided by traditional statistical tests. For instance, it is easier to understand a rule such as: "The expression level of gene A is HIGH in disease D_1 and is LOW in disease D_2 with confidence C" instead of a t-test value of 1.2. Second, fuzzy clustering algorithms are able to model genes that belong to several groups (clusters) simultaneously. This is important since a gene may have alternate roles under different conditions depending on which transcription factor regulates its expression [Gasch and Eisen, 2002]. Hence, it may be similarly expressed in more than one group of genes. This fact cannot be modeled using crisp clustering algorithms such as K-means and hierarchical clustering where a gene belongs exclusively to a single cluster. Third, fuzzy approaches account for noise in the data because they extract trends rather than precise values [Woolf and Wang, 2000]. In section 5.2 we will describe several fuzzy clustering algorithms used for finding differentially expressed genes.

Various methods for patient classification based on gene expression exist, such as neural networks, support vector machines and k-nearest neighbors [Lee, 2004]. The advantage of fuzzy classification approaches like fuzzy rule systems and the fuzzy k-nearest neighbor (FKNN) algorithm is that they are more transparent than their crisp counterparts while allowing each sample (patient) to have a certain degree of membership in each condition. At this point we would like to caution the reader when using sample (patient) classification algorithms based on gene expression. In view of the fact that training data is very limited (perhaps 100 samples) and the dimension of the feature space is very high (20,000 to 100,000 genes), one can only wonder how any classification algorithm is able to reasonably sample the search space. Obviously, the previous cautionary note will not hold if the "silver bullet" feature is found; that is, the gene that alone can differentiate between conditions. While this biomarker might exist in some genetic diseases, it does not generally appear in most of the microarray datasets where complex gene networks are involved in producing differential gene expression.

5.1.3 Microarray processing algorithms for gene regulatory network discovery

Microarrays have been used alone or in conjunction with other data sources (such as protein-protein interactions) to discover new gene regulation mechanisms. Several methods have been proposed to develop gene interaction networks including linear equations, differential equations, and Boolean networks [Ressom *et al.*, 2003]. Differential equations are the most exact method but they are not able to model a large number of variables simultaneously. Boolean networks assume that genes are either "on" or "off", a fact that contradicts biological reality. Unlike Boolean networks, fuzzy rule approaches assume different levels of transcription such as LOW, MEDIUM, and HIGH each modeled over a continuum of possible values. A fuzzy rule approach to discovering gene regulatory networks is described in section 5.3.

5.2 Clustering Algorithms

Clustering is a way of analyzing a set of objects by separating them into groups. Clustering algorithms are based on the notion of similarity. The assumption is that similar objects end up in the same group; hence, we can infer some common properties for the objects in each cluster. However, clustering of objects depends strongly on the similarity measure used to compare them. Moreover, the similarity measure (or its dual, the distance measure) determines the shape of the cluster. For example, in a two dimensional space (R^2), the Euclidean distance (L_2-norm in R^2) produces "circular" clusters while the city block distance (L_1-norm in R^2) produces "diamond" clusters. We describe more microarray similarity measures (distances) in the next section.

There are several modalities in which microarray data can be clustered. First, by clustering the gene expression profiles (rows of Table 5.1), we can discover genes co-regulated (up or down) in a certain group of samples. In Figure 5.3 a group of chromatin related genes were found after gene clustering to be under-expressed in several sample groups and over-expressed in others (circled). As mentioned previously, if a gene is regulated by several transcription factors depending on the condition of the organism, it reasonably belongs to several clusters simultaneously [Gasch and Eisen, 2002]. Such a case requires a fuzzy approach such as the fuzzy C-means (FCM) or, even more suitable, the possibilistic C-means (PCM). We will discuss the differences between the FCM and PCM in the Section 5.2.2.

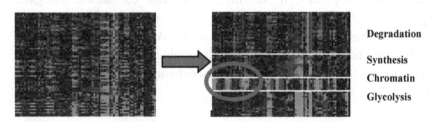

Figure 5.3 Sample microarray data before (left) and after (right) gene clustering (red=over-expressed gene, green=under-expressed gene, black=similar expression in both condition). After clustering, a group of chromatin related genes were found to be under-expressed in several groups of samples and over-expressed in others (circled).

The second way in which the data can be grouped is by clustering the samples (patients, columns of Table 5.1). This might be necessary when subgroups of a certain condition (disease) need to be discovered. In addition, the sample dimension is used when a classifier is trained to recognize the condition of a patient based on the gene expression data.

The third way of grouping the data shown in Table 5.1 is to simultaneously cluster the rows (genes) and columns (samples, patients). This clustering approach is known as co-clustering (a.k.a. two-way-clustering or bi-clustering). By co-clustering, one tries to identify groups of genes associated with group of samples (patients). In this case, we could discover the genes responsible for a certain sub-condition (for example, a subgroup of follicular lymphoma patients that respond well to treatment). As opposed to crisp co-clustering algorithms [Cheng and Church, 2000] where a gene and a patient must belong only to a given co-cluster, in fuzzy co-clustering they can both belong to multiple co-clusters simultaneously. A fuzzy co-clustering algorithm will be presented in Section 5.2.4.

The clustering of genes or samples can be performed in two basic ways, which result in two classes of clustering algorithms - object data and relational data approaches. In the object data approach, the genes are clustered based on their expression profiles (see Figure 5.2). Each gene i is represented by a point $\mathbf{x}_i \in R^M$, where M is the number of samples. In this case, the input of the clustering procedure consists of the object data $\{\mathbf{x}_i\}_{i=1,N}$. The best known object data fuzzy clustering method is fuzzy C-means (FCM). The FCM algorithm was introduced in Section 2.6.1. In Section 5.2.2 we present more details related to the application of FCM to microarray data.

A second clustering approach utilizes either a distance or dissimilarity matrix between the genes or patients. Because these methods only use the dissimilarity between the objects, they are called relational approaches. A relational clustering algorithm called Non-Euclidean Relational Fuzzy C-means (NERF C-means) [Hathaway and Bezdek, 1994] will be presented in Section 5.2.3.

As mentioned in the beginning of this section, the type of similarity (distance) measure employed in clustering greatly influences the results.

In the next section we give several similarity (dissimilarity) measures used for microarray clustering.

5.2.1 (Dis)similarity measures for microarray data

Let $\mathbf{x}_i=(x_{i1} \ldots x_{iM})$ and $\mathbf{x}_j=(x_{j1} \ldots x_{jM})$ be the expression profiles of gene i and j, respectively. There are many possible choices for calculating the similarity between two gene profiles [Xu and Wunsch, 2005]. In what follows, we discuss gene profile similarity but note that sample (patient) profiles could be used interchangeably.

The correlation (cosine) similarity between gene i and gene j is defined as [Gasch and Eisen, 2002]:

$$s_c(\mathbf{x}_i,\mathbf{x}_j) = \frac{\sum_{k=1}^{M} x_{ik} x_{jk}}{\sqrt{(\sum_{k=1}^{M} x_{ik}^2)(\sum_{k=1}^{M} x_{jk}^2)}}. \tag{5.1}$$

The correlation similarity captures the shape similarity between the two expression profiles regardless of the magnitude of the expression levels. In addition, this similarity ignores the dependence of the gene correlation on the absolute expression level. That is, the correlation between two highly expressed genes should be significant while the same value obtained between two poorly expressed genes should not be significant [Lee 2004]. The correlation similarity measure is suitable when one is interested in comparing the shape of the expression profiles as a result of changes in regulation mechanisms [Gasch and Eisen, 2002]. The correlation dissimilarity, d_c, is obtained by $d_c=1-s_c$.

Another version of the above similarity is the Pearson correlation, in which the mean expression value is subtracted from each profile. The Pearson correlation is defined as:

$$s_p(\mathbf{x}_i,\mathbf{x}_j) = \frac{\sum_{k=1}^{M} (x_{ik} - \mu_i)(x_{jk} - \mu_j)}{\sqrt{\sum_{k=1}^{M} (x_{ik} - \mu_i)^2 (\sum_{k=1}^{M} x_{jk} - \mu_j)^2}}, \tag{5.2}$$

where μ_i, μ_j are the mean values of the expression profiles x_i, x_j, respectively.

The Spearman correlation is similar to the Pearson correlation except that is based on the ranks (obtained by sorting) of the expression values in a profile instead of the values themselves. This makes the Spearman correlation less sensitive to outliers. The Spearman correlation similarity between two gene expression profiles x_i and x_j can be calculated as:

$$s_s(\mathbf{x}_i, \mathbf{x}_j) = \frac{\sum_{k=1}^{M} r_{ik} r_{jk}}{\sqrt{(\sum_{k=1}^{M} r_{ik}^2)(\sum_{k=1}^{M} r_{jk}^2)}}, \qquad (5.3)$$

where r_{ik} and r_{jk} are the rank of the k-th element of x_i and x_j, respectively.

Euclidean distance was also used in clustering gene expression profiles [Dembele *et al.* 2003, Belacel *et al.* 2004]. The Euclidean distance (see also Section 2.6) between two expression profiles is:

$$d_e(\mathbf{x}_i, \mathbf{x}_j) = \sqrt{(\mathbf{x}_i - \mathbf{x}_j)^t (\mathbf{x}_i - \mathbf{x}_j)} = \sqrt{\sum_{k=1}^{M} (x_{ik} - x_{jk})^2}. \qquad (5.4)$$

The Euclidean similarity is computed as $s_e = 1/(1 + d_e)$. Since the Euclidean distance is scale dependent, the expression profiles might require normalization such as $\{(x_{ik} - \mu_i)/\sigma_i\}_{k=1,m}$, where μ_i and σ_i are the mean and standard deviation of the expression profile x_i, respectively. Euclidean distance is the most utilized measure for comparing expression profiles for both genes and samples (patients).

Euclidean distance assumes that the gene profiles are uncorrelated, resulting in spherical shaped clusters. To obtain ellipsoidal clusters, we use the Mahalanobis distance (see also Equation 2.10) that is defined as:

$$d_m(\mathbf{x}_i, \mathbf{x}_j) = \sqrt{(\mathbf{x}_i - \mathbf{x}_j)^t \Sigma^{-1}(\mathbf{x}_i - \mathbf{x}_j)}, \qquad (5.5)$$

where Σ^{-1} is the inverse of the covariance matrix of the data set. The covariance matrix, Σ, is computed (see also Equation 2.11) as:

$$\Sigma = \frac{1}{N} \sum_{k=1}^{N} (\mathbf{x}_k - \boldsymbol{\mu})(\mathbf{x}_k - \boldsymbol{\mu})^t, \qquad (5.6)$$

where μ is the average gene expression profile, that is,

$$\mu = \frac{1}{N}\sum_{k=1}^{N}\mathbf{x}_k .\qquad(5.7)$$

The city block distance [Lee 2004] is defined as:

$$d_b(\mathbf{x}_i,\mathbf{x}_j) = \sum_{k=1}^{M}|\ x_{ik} - x_{jk}\ | .\qquad(5.8)$$

Example 5.1 Similarity calculation for the GD30 data

We generated simulated gene expression profiles for the GD30 genes (N=30) used in Chapter 3. Each profile is 20 samples (M=20) long and is hypothetically divided in two groups: "normal" patients (first 10 samples) and "cancer" patients (last 10 samples). The profiles were generate such that

- the anti-apoptotic genes (first 10 genes) have a low expression in the "normal" patient group and have a high expression in the "cancer" group;

- the pro-apoptotic genes (genes 11 to 20) have a high expression in the "normal" patient group and have a low expression in the "cancer" group;

- the other apoptosis related group (genes 21 to 30) have a medium expression in both patient groups.

The resulting simulated microarray "chip" is shown in Figure 5.4. The data will be further denoted as GD30.

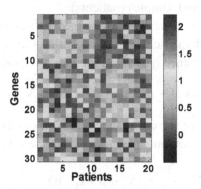

Figure 5.4 The simulated microarray data (GD30) used throughout this chapter.

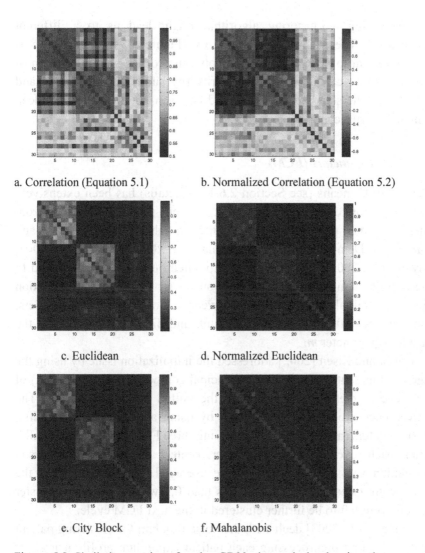

a. Correlation (Equation 5.1)   b. Normalized Correlation (Equation 5.2)

c. Euclidean   d. Normalized Euclidean

e. City Block   f. Mahalanobis

Figure 5.5 Similarity matrices for the GD30 data calculated using the measures mentioned in this section.

In Figure 5.5 we show the similarity matrices calculated with the measures mentioned in this section. From visual inspection of the similarity matrices, we are inclined to say that the correlation similarity differentiates the three types of gene profiles better than the Euclidean similarity. Aside from display considerations, we will see in Example

5.3 that using a clustering algorithm might lead us to a different conclusion. The Mahalanobis similarity (Figure 5.5.f - Equation 5.5) does not seem to differentiate at all between profiles. Normalization of profiles before calculating the similarity (by subtracting the mean and dividing by the standard deviation) did not seem to affect the Euclidean similarity significantly.

5.2.2 Fuzzy C-means (FCM)

The Fuzzy C-means (see Section 2.6.1 for details) has been extensively employed for clustering gene expression profiles obtained from microarray data [Gasch and Eisen, 2002; Dougherty *et al.*, 2002; Wang *et al.* 2003; Dembele *et al.* 2003; Arima *et al.* 2003; Belacel *et al.* 2004; Asyali *et al.* 2005]. Xu *et.al.* [2005] pointed out some issues related to successfully employing the FCM such as algorithm initialization (defining initial partition), sensitiveness to noise and outliers, convergence often to a local minimum, and difficulty in choosing the fuzziness parameter m.

Gasch and Eisen [2002] addressed the initialization issue by using the eigenvectors that resulted from a principal component analysis (PCA) of the gene expression data among the initial cluster centers. The convergence problem was addressed by using a high number of initial cluster centers (around 200) and applying three FCM cycles sequentially. After each cycle, duplicate cluster centers (pairs whose Pearson correlation was greater than 0.9) are averaged. In addition, only the genes with a Pearson correlation less than 0.7 within any of the cluster centers were left to be further clustered in the next FCM cycle.

Wang *et al.* [2003] dealt with the noise problem for clustering patient profiles by first preprocessing each patient expression profile using self organizing feature maps. Consequently, each patient expression profile was reduced to an $n_1 \times n_2$ feature map (a total of 50 to 100 nodes), resulting in M such feature maps. Next, FCM was applied on the M-dimensional feature vectors in order to identify clusters of genes that mapped into adjacent areas of the SOFM.

Belacel *et al.* [2004] employed two heuristics, called J-means and variable neighborhood search (VNS), to address the convergence of the regular FCM to local minima. In the J-means heuristics, the new centroids are chosen from the data points that are in the neighborhood of the current solution. The VNS, on the other side, is a metaheuristics that searches for more distant solutions, possible better than the local ones provided by the J-means. The J-means heuristics uses a reformulated objective function that depends only on the cluster centers (centroids) as:

$$\min_{\{C_k\}} R_m = \sum_{i=1}^{n} \left[\sum_{k=1}^{K} d(X_i, C_k)^{2(1-m)} \right]^{(1-m)}. \tag{5.9}$$

The pseudo-code for the fuzzy J-means algorithm is given below.

BEGIN
    -Choose K of the N gene profiles $\{x_i\}$ as cluster centers, $C_{opt} = \{c_k\}_{k=1,K}$;
    -Compute $R_{opt}^{old} = R_m(C_{opt})$ using (5.9) and set $R_{opt}^{old} = 10^{10}$;
    -Choose:
        -stopping constant ε,
        -maximum search neighborhood size q_{max};
    WHILE $(R_{opt}^{old} - R_{opt}^{new}) > \varepsilon$ DO

        $q=1$, $R_{opt}^{old} = R_{opt}^{new}$.

        WHILE $q < q_{max}$ DO
        -Replace at random q new centroids from the unassigned $\{x_i\}$
        -Drop least useful centroid c_d (that produces the greatest
          increase in R_m);
        -Add most useful available pattern x_u (that produces the
          smallest increase in R_m) as centroid;
        -Recalculate $R_{opt}^{new} = R_m$;

        -IF $(R_{opt}^{old} - R_{opt}^{new}) > \varepsilon$
          reset q=1;
          $R_{opt}^{old} = R_{opt}^{new}$;
        ELSE
          $q=q+1$;
        END IF
        END WHILE

END WHILE
-Compute fuzzy memberships u_{jk} using (2.9)
END

Dembele *et al.* [2003] proposed calculating the fuzziness parameter as a function of the coefficient of variation of the set of distances between the gene expression profiles, $D = \{d_{ij}^{2/(m-1)} \mid i \neq j; i, j \in [1, N]\}$. The coefficient m is computed by numerically solving the equation $\sigma_D / \mu_D = 0.03m$, where σ_D and μ_D are the standard deviation and the mean of the set D, respectively.

As in any clustering algorithm, one problem that requires special attention is noise and outlier handling. In FCM, the outlier handling concerns are raised by the constraint (Equation 2.6) that the memberships in clusters for each point should sum to 1. Several solutions to this problem have been proposed in the literature [Dave, 1991; Krishnapuram and Keller, 1993]. Dave [1991] used an extra noise cluster represented by a prototype (cluster center) that has the same distance, δ, to any data point. This choice is based on the assumption that outliers are equidistant to the center clusters (fact that might not be true in all cases). In this case the cluster memberships (Equation 2.9) become:

$$u_{jk} = \cfrac{1}{\displaystyle\sum_{i=1}^{C}\left[\frac{d(\mathbf{x}_k, \mathbf{v}_j)}{d(\mathbf{x}_k, \mathbf{v}_i)}\right]^{\frac{2}{m-1}} + \left[\frac{d(\mathbf{x}_k, \mathbf{v}_j)}{\delta}\right]^{\frac{2}{m-1}}}, \qquad (5.10)$$

where the noise distance δ is computed as:

$$\delta = \sqrt{\lambda \frac{\displaystyle\sum_{k=1}^{N}\sum_{i=1}^{C} d^2{}_{ik}}{NC}}, \qquad (5.11)$$

with λ a scale parameter.

Another possible solution was given by the possibilistic C-means algorithm (PCM, Section 2.6.3) [Krishnapuram and Keller, 1993]. PCM removes the membership summation constraint (Equation 2.6) and allows each object to belong to any of the clusters with a membership $u_{ij} \in [0,1]$. Aside from dealing with noise, this seems to be the natural solution for many bioinformatics problems. Following the example given in [Gasch and Eisen, 2002], if a gene is controlled by several transcription factors then it should have a high membership (say, equal to 1) in any of the corresponding clusters and a low one in all of the others. A similar interpretation is valid for the case of clustering of genes in families. A gene should be able to belong to several families (clusters) with membership 1, while another gene might not belong to any family represented in the data set (membership close to 0 in all clusters).

Example 5.2 FCM on the GD30 data.

We applied FCM with C=3 clusters on a slightly modified version of the GD30 data from Example 5.1. In order to demonstrate the danger of outliers in FCM, we modified the profile of gene 21 (the first gene from the "other apoptosis related genes" group, class 3) by setting it to a "very high" expression in both "normal" and "cancer" groups. The modified gene profile is clearly an outlier, as can be seen in Figure 5.6 of the 2-dimensional representation of the data obtained by applying principal component analysis (pointed at with an arrow in Figure 5.6).

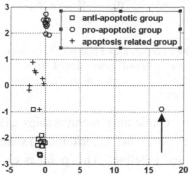

Figure 5.6 The GD30 was represented in 2 dimensions by performing principal component analysis and keeping only the first 2 components. The outlier gene is indicated by arrow.

The cluster memberships produced by FCM are shown in Figure 5.7. We see that the membership of the outlier in the pro-apoptotic group (dotted arrow) was slightly higher (0.4) than in the anti-apoptotic and apoptosis related groups (0.3 in each). As a consequence, the outlier would be erroneously assigned to class 2 when the membership array is hardened (cluster assignment performed based on maximum membership). One of the greatest advantage of FCM over the hard C-means, is that the former allows us to tell that the assignment of this outlier was a close call. One easy way to use this information during the cluster assignment step (membership hardening) is to avoid to assign a point to clusters (hence labeling it as "outlier") if its maximum membership does not exceed a given threshold (say, 0.5 in our case).

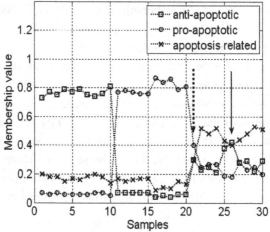

Figure 5.7 The memberships produced by FCM with *C*=3 on GD30 data. We see that the first gene in class 3 ("apoptosis related" genes, dotted arrow) was erroneously assigned to class 2 ("pro-apoptotic" genes). The sixth gene in class 3 (continuous arrow) was also erroneously assigned to class 1.

Moreover, having continuous memberships allows us to pass the "closeness" information to other algorithms downstream that are part of the same computational pipeline. This point is clearly demonstrated in Section 2.6.1.

5.2.3 Relational fuzzy C-means

Several fuzzy relational algorithms have been described in the literature such as fuzzy C-medoids [Krishnapuram *et al.*, 2001], relational fuzzy clustering [Dave *et al.*, 2002] and relational fuzzy C-means [Bezdek *et al.*, 1999]. Out of the many versions of the relational fuzzy C-means

(RFCM) algorithms presented in [Bezdek *et al.*, 1999] we will further describe the non-Euclidean relational fuzzy C-means (NERFCM) algorithm [Hathaway *et al.*, 1994], a general method applicable to numerous situations.

Let D be the matrix of dissimilarities between a set of objects. For instance, for the set of samples (patients) $\{\mathbf{p}_j\}_{j=1,M}$ the distance matrix is $D = \{d_{ij} \mid i, j \in [1, M]\}$ where d_{ij}=dist($\mathbf{p}_i, \mathbf{p}_j$) and "dist" is one of the distances mentioned in section 5.2.1. If D is computed starting from the object data using some distance measure (as in the previous example), then D is called Euclidean (that is, the triangle inequality is satisfied). If D is computed using some similarity measure (such as computing sequence similarity using BLAST) then it might not be Euclidean; that is, one might not find M vectors whose pair-wise distance matrix equals D. We mention that the problem of finding object data given their dissimilarity matrix is addressed by multi-dimensional scaling (MDS) methods [Cox and Cox, 2001]. MDS could be used in an alternative approach to clustering relational data; that is, after finding the objects whose dissimilarity matrix is D, one can apply any object data fuzzy clustering algorithm.

In order for NERFCM to work, D has to satisfy three conditions:

    1. d_{jj}=0, for all $j \in [1,M]$,

    2. $d_{jk} \geq 0$, for all $j,k \in [1,M]$,

    3. d_{jk}=d_{kj}, for all $j,k \in [1,M]$.

NERFCM is an iterative algorithm. In each iteration, the dissimilarity matrix D is first readjusted using a *β-spread* transform computed using the average membership in a cluster, $\mathbf{v}_i$, i=1,C. Then, the membership matrix U is recomputed based on the adjusted dissimilarities. The detailed steps of the NERCM algorithm are as follows:

BEGIN

    -Choose number of clusters $C \in [2,M)$, fuzzifier m>1 and stopping
        threshold ε.

    -Initialize the cluster membership matrix U=$\{u_{ik}\}$, $i \in [1,C]$,
    $k \in [1,M]$ and parameter β=0.

    -DO UNTIL $\left\| U^{(old)} - U^{(new)} \right\| \leq \varepsilon$

FOR each $i \in [1, C]$, compute the "C-means" vectors as:

$$\mathbf{v}_i = ((u_{i1})^m, (u_{i2})^m, \ldots, (u_{iM})^m) \Big/ \sum_{k=1}^{M} (u_{ik})^m , \; i \in [1, C],$$

END FOR i

FOR each cluster i

   FOR each point k

      -Re-compute the distances d_{ik}:

$$d_{ik} = (D \mathbf{v}_i)_k - (\mathbf{v}_i^t D \mathbf{v}_i)/2$$

      -IF $d_{ik} < 0$ THEN matrix D is not Euclidean

   END FOR k

END FOR i

IF D is not Euclidean

      - adjust parameter β as: $\beta = \beta + \Delta\beta$

$$\text{where } \Delta\beta = \max_{i,k} \left\{ -2 d_{ik} \big/ \|\mathbf{v}_i - \mathbf{e}_k\|^2 \right\}$$

$$\text{and } \mathbf{e}_k = (0, \ldots, \underbrace{1}_{k}, \ldots, 0)^t \in R^M .$$

      - Re-computed the d_{ik} as:

$$d_{ik} = d_{ik} + (\Delta\beta/2) \cdot \|\mathbf{v}_i - \mathbf{e}_k\|^2 .$$

END IF

FOR each cluster i

   FOR each point k

      Re-compute the cluster memberships $\{u_{ik}\}$ as:

$$u_{ik} = \begin{cases} \left(\sum_{j=1}^{C} \left(d_{ik} / d_{jk} \right)^{\frac{1}{m-1}} \right)^{-1} & \text{if} \quad d_{ik} \neq 0 \\ 0 & \text{if} \quad d_{ik} = 0 \end{cases}$$

   END FOR k

  END FOR i

 END DO UNTIL

END

Example 5.3 NERFCM of the GD30 data.

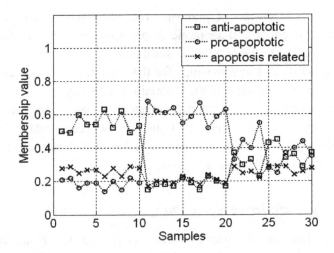

Figure 5.8 The NERFCM cluster memberships for the correlation dissimilarity. All the genes from the "apoptosis related" group (samples 20 to 30) were wrongly clustered.

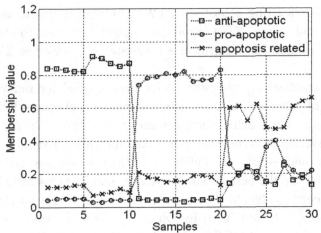

Figure 5.9 The NERFCM cluster memberships for the Euclidean similarity. All the genes from the "apoptosis related" group (x-marker) were correctly clustered.

For this example we used the correlation similarity for GD30 profiles (no outlier, see Figure 5.5.a) to obtain a dissimilarity matrix. We ran NERFCM with $C = 3$, $m = 1.5$, $\varepsilon = 0.001$. The resulting cluster memberships for the correlation similarity are shown in Figure 5.8. While the first two classes were clustered in the expected cluster, the "noise" class elements (apoptosis related genes) were clustered erroneously as either anti-apoptotic or pro-apoptotic. Is this due to the NERFCM or due to the correlation similarity? If we analyze Figure 5.9, where the NERFCM memberships for the Euclidean distance are given, we see the problem resides in the correlation similarity.

Both object and relational clustering are very useful in bioinformatics. Since the relationships between data, clustering algorithm and dissimilarity are complex, care must be taken to not "read too much" into the output of any given choice.

5.2.4 Fuzzy co-clustering algorithms

Many co-clustering techniques have been used for biological data analysis [Madeira and Oliveira 2004]. Cheng and Church [2000] employed a mean squared residue method to cluster a yeast cell cycle microarray data set. Getz *et al.* [2000] used an alternate row-columns hierarchical clustering algorithm, CTWC, to analyze microarray data. Lazzeroni and Owen [2000] introduced a plaid model for co-clustering where the value of an element in a co-cluster is modeled as a sum of layers.

Fuzzy co-clustering algorithms have been applied for simultaneous clustering of documents and words in text mining applications [Oh *et al.*, 2001; Frigui *et al.*, 2002; Kummamuru *et al.*, 2003; Tjhi and Chen, 2006]. Aside from normalization and data dimension issues, clustering documents and words is similar to clustering patients and genes, respectively. To the best of our knowledge, fuzzy co-clustering has not been applied so far to microarray data. The fuzzy co-clustering algorithm, FCCM, introduced by Oh *et al.* [2001] is described in the remainder of the section, followed by an example on our GD30 dataset.

We define constraints on the memberships of a patient i and a gene j to a co-cluster c as:

$$\sum_{c=1}^{C} u_{ci} = 1, \quad u_{ci} \in [0,1], \quad i = 1,...,M, \quad (5.12)$$

$$\sum_{j=1}^{N} w_{cj} = 1, \qquad w_{cj} \in [0,1], \quad c = 1,...,C \qquad (5.13)$$

where u_{ci}, w_{cj} are the memberships of the ith patient and of the jth gene in the cth co-cluster, respectively. The objective function for FCCM can be written as:

$$\max\{ J = \sum_{c=1}^{C}\sum_{i=1}^{M}\sum_{j=1}^{N} u_{ci} w_{cj} x_{ij}$$

$$-T_u \sum_{c=1}^{C}\sum_{i=1}^{M} u_{ci} \log u_{ci} - T_w \sum_{c=1}^{C}\sum_{j=1}^{N} w_{cj} \log w_{cj} \}, \qquad (5.14)$$

where T_u and T_w are weighting parameters which specify the degree of fuzziness. By applying the Lagrange multipliers method to the above objective function, we find the necessary condition for the patient i and gene j memberships in the co-cluster c as:

$$u_{ci} = \frac{\exp(\sum_{j=1}^{N} w_{cj} x_{ij} / T_u)}{\sum_{c=1}^{C} \exp(\sum_{j=1}^{N} w_{cj} x_{ij} / T_u)}, \qquad (5.15)$$

$$w_{cj} = \frac{\exp(\sum_{i=1}^{M} u_{ci} x_{ij} / T_w)}{\sum_{j=1}^{N} \exp(\sum_{i=1}^{M} u_{ci} x_{ij} / T_w)}. \qquad (5.16)$$

In order to avoid overflows, for large N, care should be exercised when computing the gene memberships, w_{cj}. One solution is to reduce N by prescreening the genes before applying FCCM. If the number of genes is still large, we can use alternative algorithms such as Fuzzy-CoDoK [Kummamuru *et al.*, 2003] or FCC-STF [Tjhi and Chen, 2006] that reportedly do not exhibit this problem. The pseudo-code for the FCCM algorithm is:

BEGIN
    -Choose C, T_u, T_w and ε.

-DO UNTIL $\max | u_{ci}^{NEW} - u_{ci}^{OLD} | < \varepsilon$
    FOR each cluster $c=1,C$
        -FOR each patient $i=1,N$
            Update patient memberships using u_{ci} (5.15)
        END FOR i
        -FOR each gene $j=1,M$
            Update gene memberships w_{cj} using (5.16)
        END FOR j
    END FOR c
  END DO
END

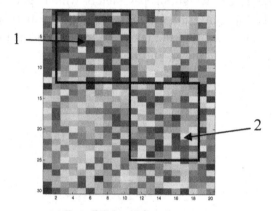

Figure 5.10 The rearranged GD30 data as a result of applying FCCM with $C=3$ on the GD30 data. Two co-clusters (marked with "1" and "2") were identified.

Example 5.4 Fuzzy Co-clustering of the GD30 data

We co-cluster the GD30 profiles (no outlier) with the FCCM algorithm. We used the following values for initialization: $C=3$, $T_u=1$, $T_v=T_u N/M$ and $\varepsilon=0.00001$. After convergence of FCCM, we performed a gene assignment to the three co-clusters. The cluster assignment of gene j was obtained by finding $argmax_{c=1,C}\{w_{cj}\}$ resulting in the following gene (rows) assignment {1, 1, 1, 1, 1, 1, 1, 1, 1, 1, 2, 2, 2, 2, 2, 2, 2, 2, 2, 2, 2, 1, 1, 1, 2, 2, 2, 3, 1, 2}. Similarly, patients were assigned to co-clusters. The cluster assignment of patient i was obtained by finding $argmax_{c=1,C}\{u_{ci}\}$. The resulting assignment for the patients (columns) was: {2, 2, 2, 2, 2, 2, 2, 2, 2, 2, 1, 1, 1, 1, 1, 1, 1, 1, 1, 1}. Using the above assignments to re-arrange the rows and columns we obtain the co-clusters shown in Figure 5.10.

It is interesting to note in Figure 5.10 that third co-cluster does not have any column assignments although it has a gene assignment (last row in Figure 5.10, which was gene number 28 in the initial GD30).

5.3 Inferring Gene Networks Using Fuzzy Rule Systems

The genes active in a given cell interact to form a complex regulatory network. The gene regulatory network (GRN) dynamically establishes the level of expression of each gene as a function of the state of the cell (internal conditions) and of the environment (external conditions). That is, the expression of each gene at a given time is a function of internal (expression of other genes) and external conditions. As a consequence, theoretically, one can use the gene expression levels to infer, by reverse engineering, the connection between genes for different cell types (normal or abnormal) or different environmental conditions (i.e. day or night). While theoretically this approach is appealing, it presents two major challenges [Guthke *et al.*, 2005]. First, microarray data is noisy, hence the inferred connections are not necessarily reliable . Second, the data is insufficient for inferring all existing connections, since a $N \times N$ interaction matrix mast be estimated using only $N \times M$ data points, where $M << N$.

Fuzzy logic provides the mathematical framework to better address the first challenge than do statistical methods, due to its ability to qualitatively deal with noisy data [Sokhansanj *et al.*, 2004]. As we described in Section 2.5, instead of using a real number, a fuzzy rule system typically uses three levels to describe the expression value: LOW, MEDIUM and HIGH. As a consequence, the variation present in the data due to noise is ameliorated. Among the approaches that used fuzzy rule systems to infer GRNs, we mention [Woolf and Wang, 2000; Ressom *et al.*, 2003; Sokhansanj *et al.*, 2004]. Woolf and Wang used an intersection fuzzy rule system (IFRS) to find activator-repressor-target triplets in yeast cell cycle microarray data. An IFRS is a fuzzy rule system with rules of the form "**IF** (expression of) A is High and (expression of) B is Low **THEN** (expression of) C is High". Ressom *et al.* improved the speed of the previous algorithm by 50% by preprocessing the data using a SOFM. The problem with this approach is that IFRS is limited to two or three inputs due to the combinatorial explosion of the number of rules. To address this difficulty, Sokhansanj *et al.* used a union fuzzy rule system (UFRS) to determine the relationships between 12 yeast cell cycle genes. An UFRS will break the above intersection rule in two: "**IF** A is High **THEN** C is High" and "**IF** B is Low **THEN** C is High". While this approach is simpler, it assumes that the inputs, A and B, are essentially independent (noninteractive). Reportedly, this might not be a bad assumption given the fact that nature usually favors simple mechanisms (Occam's razor) [Guthke *et al.*, 2005].

To address the second challenge, bioinformaticians usually employ some dimensionality reduction technique such as fuzzy clustering [Gutke *et al.*, 2005; Sehgal *et al.*, 2006; Du *et al.*, 2005] or self organizing maps (SOM) [Ressom *et al.*, 2003]. Gutke *et al.* employed FCM to cluster the gene expressions and then they used differential equations to model the cluster centers. After clustering the gene expressions of the plant *Arabidopsis Thaliana* using FCM, Du *et al.* computed the correlation coefficient between the cluster representatives. A positive correlation was interpreted as activation and a negative one as inhibition. These approaches generally do not work in real applications except for some very simple cases. To address the second challenge, the fuzzy logic framework provides a different solution in which no training data is necessary. Instead, we use domain expert knowledge (gene regulation dynamics, in this case) to set up a fuzzy rule system. The knowledge (gene relationships) is then inferred from the existent data using the fuzzy rule system.

As an example, we present the work of Woolf and Wang [2000] who employed an IFRS of the Mamdani type (see Section 2.5) to find {activator (*A*), repressor (*R*), target (*T*)} triplets in microarray expression data. A gene *A* is an activator for gene *T* if an increase in the expression of *A* induces an increase in the expression of *T* and vice-versa. A repressor gene *R* has the opposite effect on *T*. To model the coupled effect of an activator and a repressor on a target, Woolf and Wang used the following fuzzy rule system:

IF A=High and R=High **THEN** T=Medium
IF A=High and R= Medium **THEN** T=High
IF A=High and R= Low **THEN** T=High
IF A=Medium and R=High **THEN** T=Low
IF A=Medium and R= Medium **Then** T=Medium
IF A=Medium and R= Low **Then** T=High
IF A=Low and R=High **Then** T=Low
IF A=Low and R= Medium **Then** T=Low
IF A=Low and R= Low **Then** T=Medium.

The membership functions used by Woolf are as follows:

$$Low = \begin{cases} -x & x \in [0,0.5] \\ 0 & x \in (0.5,1] \end{cases},$$

$$Medium = \begin{cases} x & x \in [0,0.5] \\ -x & x \in (0.5,1] \end{cases}, \tag{5.17}$$

$$High = \begin{cases} 0 & x \in [0,0.5] \\ x & x \in (0.5,1] \end{cases}.$$

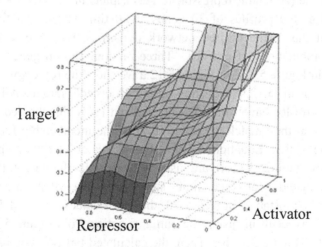

Figure 5.11 The control surface for the fuzzy rule system with two inputs (activator-A, repressor-R) and one output (target-T) [Woolf and Wang, 2000].

The resulting graph of running this IFRS over all activator/repressor pairs FRS is shown in Figure 5.11.

Note that the fuzzy rule system with only 9 rules provides a smooth output function, addressing the issue of noise in microarray data.

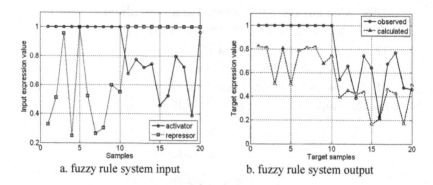

a. fuzzy rule system input   b. fuzzy rule system output

Figure 5.12 An example of (activator, repressor, target) found in the GD30 data at a 10% relative squared error threshold; in a. the inputs of the FRS are shown; in b. the FRS output (triangles) is compared to the observed gene expression profile (circles).

Example 5.5 Finding (activator, repressor, target) in the GD30 data

To find the (activator, repressor, target) triplets in the GD30 data we interpret it as the profiles of 30 genes at 20 time steps. To find the triplets that match the desired network dynamics, we have to try all possible combinations of profiles. Three random chosen gene profiles are shown in Figure 5.12. In Figure 5.12.a we show the repressor and the activator and in 5.12.b we show the "observed" target. All three expression profiles have to be first normalized (only thresholded in our case) such that they match the range [0,1] of the membership functions used in the IFRS (Equation 5.16). The activator and the repressor profiles (see Figure 5.12.a) are used to calculate the target (Figure 5.12.b-"calculated") at each time step using the IFRS previously described. For instance, at step $t=10$, the input values $A=1$ and $R=0.55$ (Figure 5.12.a) result in (using the output function from Figure 5.10) an output $T=0.7$ (Figure 5.12.b). Then, the calculated target is compared to the observed one (see Figure 5.12.b). If the relative square error between the two target profiles is smaller than a certain threshold (10% in our case), the triplet is recorded as a possible (activator, repressor, target) network. An example of such a network is shown in Figure 5.12.

5.4. Discussion and Summary

In this chapter we presented several fuzzy techniques for processing of microarray data. Fuzzy clustering is, by far, the most important technique of all, due to its property of multiple cluster memberships that matches the biological foundations of gene expression. We have shown that although useful, all clustering and fuzzy clustering in particular, has its pitfalls such as noise sensitivity and dependence on the distance measure employed. When using a clustering algorithm, we have to understand its intricacies since there is no such thing as the "best clustering algorithm" but only a "suitable for the data" one.

In the end, we have shown how fuzzy rule systems can be used in inferring gene regulating networks. A fuzzy rule system represents a modality to use expert knowledge in finding patterns in data.

Chapter 6

Other Applications

6.1 Overview

Fuzzy set theory and fuzzy logic, like many other computational intelligence methods, can be used for a wide range of bioinformatics problems. In this book, we have addressed applications of fuzzy approaches in ontology, protein structure prediction, and microarray data analysis. In this chapter, we will summarize some other applications.

We will focus on the applications on biological sequence analyses, computational proteomics, drug design, and biomedical text mining. Each of these four areas represents important bioinformatics research and has significantly utilized fuzzy set theory and fuzzy logic. We will also briefly touch a few other areas, where a limited number of fuzzy approaches have been deployed.

There are a number of pertinent reviews. The paper by Torres and Nieto [2006] provides an overview on the application of fuzzy logic in medical informatics and bioinformatics. It shows two examples with some details, i.e., drug addiction and genome comparison. Another review by Mitra and Hayashi [2006] has a brief summary on the application of fuzzy sets in bioinformatics, together with applications of other soft-computing techniques such as artificial neural networks and evolutionary computing. The authors of this book also briefly reviewed the applications of fuzzy logic in bioinformatics [Xu *et al.*, 2006]. Nevertheless, none of the previous reviews has the breath of depth on the subject like this book.

6.2 Applications in Biological Sequence Analyses

Biological sequences, as described in Appendix I, are the most fundamental objects for biological systems at the molecular level. Biological sequences range from a gene sequence (in the form of DNA, RNA, or protein) to the whole genome. A gene sequence encodes all the information related to the structure and function of the gene product (protein). The genome sequence is also called the blueprint of life and it defines a biological species. Thanks to the Human Genome Project and various other genome projects, massive biological sequence data have been generated and are being produced at an exponential rate. Hundreds of genomes have been completely sequenced, including human, mouse, rice, worm, and many bacteria. It is projected that more than 10,000 genomes will be sequenced in a decade. Various sequencing efforts have also produced a tremendous number of sequences of DNA segments and proteins.

Given vast amounts of biological sequences available, it is impossible to characterize all of them experimentally. Computationally analyzing these sequences provides valuable information for understanding the biological systems. Other computational intelligence techniques, such as artificial neural networks and support vector machines have been widely applied in biological sequence analyses. Meanwhile, fuzzy set theory and fuzzy logic have also started gaining ground in examining biological sequences.

6.2.1 Protein sequence comparison

Comparison of biological sequences, including protein sequences and nucleotide sequences (DNA and RNA), is the most fundamental technique in bioinformatics. Comparing sequences can establish evolutionary relationship between different bio-molecules or different biological species, and as a result, infer biological structures and functions. Sequence comparison also offers a basis for new medical diagnosis and drug development, as well as bioengineering of modified species. It is one of the most intensive applications on many supercomputers.

The basis of sequence comparison is to match two strings with a scoring function of substitution, deletion or insertion of characters [4 types of nucleotides for DNA/RNA, i.e., A, T/U, C, G (see Section AI.1), and 20 types of amino acids, i.e., A, C, D, E, F, G, H, I, K, L, M, N, P, Q, R, S, T, V, W, and Y (see Section AI.2)]. Such a problem can be addressed with the dynamic programming technique and its variants (especially BLAST [Altschul *et al.*, 1990]).

Sequence comparison methods are developed based on the longest common substring (LCS) problem in computer science. The LCS problem can be formulated as follows: given two sequences of characters, find the longest contiguous sequence appearing in both. This problem can be solved using dynamic programming. The comparison between a sequence with length m and a sequence with length n using dynamic programming can be done in $O(mn)$ time. Although dynamic programming can compare a pair of biological sequences effectively, there are some issues in practical applications.

The first issue is that sometimes a sequence may not be uniquely identified. In particular, with high-throughput sequencing methods, a position in a DNA sequence may correspond to multiple nucleotides, each of which has a probability (or membership value). In this case, imprecise polynucleotides (words consisting of A, T, C, and G) can be formulated as fuzzy sets, i.e., points in a hypercube [Sadegh-Zadeh, 2000]. The dimension of the fuzzy set for a DNA sequence with n polynucleotides is $4n$, where each element represents the membership value for A, T, C, and G at one sequence position in the DNA. Sadegh-Zadeh [2000] introduced a simple Hamming distance to measure dissimilarities between DNA sequences. This idea was extended from polynucleotides to general biopolymers (including proteins) [Casasnovas and Rosselló, 2005].

The second issue for dynamic programming in sequence comparison is that it is time consuming for long sequences. In particular, it is not feasible to apply it to compare two genomic sequences directly. Various simplified representations for genomes have been developed. We can use statistical properties of the protein-coding regions to represent a genome. Torres and Nieto [2003] formulated a triplet codon (see Appendix I) as a 12-dimensional fuzzy set. Instead of projecting a genome sequence to a hypercube, the authors used the frequencies of the nucleotides at the three base sites of a codon in the coding sequences of a genome as membership values in a fuzzy set. Hence, each genome is represented only by a 12-dimensional vector and different genomes can be compared utilizing these vectors. The authors applied their method to the *Escherichia coli* K-12 and *Mycobacterium tuberculosis* H37Rv genomes. Different metrics for comparing the fuzzy sets between two genomes were also explored in [Nieto *et al.*, 2006].

The third issue with explicit sequence comparison as the LCS problem is that it may not be sensitive enough to detect remote relationships. Various approaches, such as hidden Markov models, attempted to code biological sequences for more sensitive detection of relationships. The signal coded in a DNA sequence can be coded in the W-curve [Cork *et al.*, 2002]. The W-curve is a numerical mapping of a DNA sequence to a profile along the sequence. It first codes the four types of nucleotides (A,C,G,T) as follows: A = (-1, 1), C = (-1,-1), G = (1, 1), and T = (1,-1). Then it maps a DNA sequence s_i, $i = 1, 2, 3...$, into a two-dimensional profile X_i, where $X_i = k (X_{i-1} + s_i)$, where $X_0 = (0,0)$ and k is a positive real number. For example, with $k=1$, one can code a sequence AATCGT as

$$(0,0), (-1, 1), (-2,2), (-1,1), (-2,0), (1,1), (2,0)$$

We can visualize the profile of X_i in a three-dimensional plot (together with the dimension i), which can be used to study the properties of a DNA sequence. Cork and Toguem [2002] formulated the W-curve as a fuzzy system. They used the power (or energy) derived from the Fourier Transform of the W-curve as the membership function. Using such a fuzzy system as a distance measure between DNA sequences improves the accuracy of the distance metric employed in sequence comparison and phylogenetic tree generation for genomic sequences. Although the fuzzy system contains less information than LSC-based approaches, it may better represent the key characteristics with less noise.

6.2.2 Application in sequence family classification

Proteins can be classified into families according to their sequence relationships derived from sequence comparison. A protein in the context of its family is much more informative than the single protein itself. For example, residues conserved across the family often indicate special functional roles. Two proteins classified in the same sequence family may suggest that they share similar structures. Different sequence comparison methods produce various ways to classify protein sequences into families and to align the members of a family.

Several sequence-based families are publicly available, including Pfam (http://pfam.wustl.edu/), ProDom (http://protein.toulouse.inra.fr/prodom/current/html/home.php), and Clusters of Orthologous Group (COG; http://www.ncbi.nlm.nih.gov/COG/new/). These methods differ in the techniques used to construct families, while they all use crisp clustering methods. However, these methods, although with significant success, do not completely solve the protein family classification problem, since the patterns in a given protein family may be too weak to detect/define. Various alternative methods are being actively explored.

For more sensitive classification of proteins, Heger and Holm [2003] developed a fuzzy scoring model for assigning a query protein sequence into one of known families. In this model, the protein sequences in a known family are pre-aligned. The authors applied multivariate analysis to define a set of attributes for the protein family, covering a subset of aligned positions that are important to the protein family. Each attribute can be defined as a sequence pattern feature such as a particular sequence position, say 210, has two possible amino acid types, e.g., D and E. This attribute has membership values (e.g., D and E with membership values 0.6 and 0.4) to the protein family. In this way, the dimensionality of sequence comparison for protein classification is reduced while the accuracy improves. The method has demonstrated a proof of principle in an extremely diverse protein family related to urease.

Since protein family classification in essence is a clustering problem for protein sequences, various fuzzy clustering methods can be applied. Bandyopadhyay [2005] extracted a set of features from the training sequences for each protein family. The extracted features are represented as the distribution of the amino acids in different positions of the aligned sequences in a protein family. Then the author applied a genetic fuzzy clustering approach to evolve a set of prototypes representing each protein family. The nearest prototype rule is used to classify an unknown sequence into a particular protein family, based on its proximity to these prototypes. Using known protein families, such as globin, trypsin and ras, the author showed the method significantly improved computing speed while providing comparable classification performance to some existing methods.

6.2.3 Application in motif identification

In a protein sequence family, some regions are better conserved than others during evolution. These regions are generally important for the function of a protein or for the maintenance of its three-dimensional structure. These regions can often be represented by motifs, i.e., short sequence segments with (nearly) conserved amino acids among related proteins. As motifs typically represent important biological functions, they can be used to assign a newly sequenced protein to a specific family, although false positive rates are high due to chance matches in short motifs. A number of motif libraries and motif-based search engines are available online, including PROSITE (http://au.expasy.org/prosite/), PRINTS (http://umber.sbs.man.ac.uk/dbbrowser/PRINTS/), BLOCKS (http://www.psc.edu/general/software/packages/blocks/blocks.htm), and the MOTIF search engine at http://motif.genome.ad.jp/ that includes PROSITE, BLOCKS, PRINTS, etc. combined.

Fuzzy methods have been used for protein motif identification. Motifs sometimes are fuzzy or flexible, i.e., the conservations of amino acids do not have to be strict. Fuzzy logic was used to describe such flexibility of protein motifs in conjunction with neural networks [Chang and Halgamuge, 2002]. A neural-fuzzy network was employed to optimize the feature of a motif, which was represented as a fuzzy rule base. The algorithm was demonstrated for proof of principle using the well known motifs C_2H_2 zinc finger and epidermal growth factor motif. In another study, researchers combined information theory with fuzzy logic search procedures to identify sequence motifs [Atchley and

Fernandes, 2005]. The method was used to identify sequence motifs for the protein family related to the Myc-Max-Mad transcription factor network.

Fuzzy methods have also been applied for DNA sequence motif identification, especially transcription factor binding sites in genomic sequences. These DNA motifs are represented by short nucleotide patterns. Genes sharing a common transcription factor binding motif may be co-regulated, resulting in a similar gene expression pattern (see Section 5.2). Pickert *et al.* [1998] proposed a fuzzy clustering approach for revealing patterns of transcription factor binding motifs. Liang *et al.* [2004] developed the cWINNOWER algorithm for fuzzy-motif detection in DNA sequences. It predicted motifs based on a clique consisting of a large number of mutated copies of the motif. The algorithm may detect much weaker signals in motifs. Cotik *et al.* [2005] proposed a hybrid analysis method to discover DNA motifs that combines neural networks, fuzzy sets, and the multi-objective evolutionary algorithms. The method was tested by learning and predicting the RNA polymerase motif in prokaryotic genomes.

6.2.4 Application in protein subcellular localization prediction

A protein usually localizes at a specific compartment in a cell (such as cell surface, cytoplasm, nucleus, etc.). Interestingly, protein localization strictly speaking is a fuzzy term, as some proteins can be trans-localized between two different compartments (especially between cytoplasm and nucleus). Protein localization information is important in understanding protein functions. A protein's subcellular location (or its primary subcellular location) can be predicted from its sequence and a wide range of computational techniques (such as neural networks, hidden Markov models, and support vector machines) have been applied to this problem.

The fuzzy k-nearest neighbors (FKNN) algorithm has also been applied to predict a protein's subcellular location [Huang and Li, 2004]. The authors derived a membership function for dipeptide composition of protein sequences in different localizations. They trained and tested their method using proteins with known localizations in the SWISS-PROT database. The overall prediction accuracy was reported about 80%. They also applied their method in predicting localizations of proteins in six genomes including human, yeast, worm, fly, rice, and plant *Arabidopsis thaliana*.

6.2.5 Genomic structure prediction

A genome has a well defined structure in its sequence. It includes coding regions and non-coding regions (see Appendix I). A gene that codes for a specific protein in eukaryotes (species whose cells contain distinct membrane-bound nuclei, e.g., animals and plants) often has a few segments for coding (called exons), separated by non-coding sequences (called introns). Gene finding is to identify introns and exons in a segment of a DNA sequence using pattern recognition algorithms. As the most important phase of genome annotation, gene finding facilitates the translation of a genomic DNA sequence into the amino acid sequence of a protein. Dozens of computer programs for identifying protein-coding genes in large genomic sequences are available. Some of the well known ones include Genscan (http://genes.mit.edu/GENSCAN.html), GeneMarkHMM (http://opal.biology.gatech.edu/GeneMark/), GRAIL (http://compbio.ornl.gov/Grail-1.3/), Genie (http://www.fruitfly.org/seq_tools/genie.html), and Glimmer (http://www.tigr.org/softlab/glimmer).

Computational gene identification from genome sequence alone (*ab initio* prediction) remains a challenging problem, especially for large-size eukaryotic genomes, as gene finding from various software packages often contain significant errors. In particular, the boundaries between exons and introns are often hard to define. To address this issue, Arredondo *et al.* [2005] developed a fuzzy inference engine based on information-theoretic considerations to predict coding regions. A set of rules were derived from known cases of exon-intron boundaries in the "if-then" format in a fuzzy way (see Chapter 2). These rules were in turn used for the prediction. The authors performed some simulated studies using human and bacterial data to illustrate the method.

Furthermore, fuzzy models were applied to study genome structure in prokaryotes. In prokaryotes (species whose cells lack of distinct membrane-bound nuclei, e.g., bacteria), each gene for coding a specific protein typically has a contiguous DNA sequence (i.e., one exon without any intron). A closely related group of neighboring genes on a DNA sequence can form an operon structure, which is regulated simultaneously. Jacob *et al.* [2005] used a fuzzy scoring function based on diverse biological information (e.g., genome sequences, functional annotations and conservation across multiple genomes) to predict operons. A genetic algorithm was employed to start from a population of putative operons in a genome into progressively better predictions. The method was tested on *Escherchia coli* K12 and *Bacillus subtilis* with good performance.

6.3 Application in Computational Proteomics

Proteomics is a leading technology for the qualitative and quantitative characterization of proteins and their interactions on a genome scale. The objectives of proteomics include the identification of their primary amino-acid sequence, large-scale identification and quantification of all protein types in a cell or tissue, analysis of post-translational modification and association with other proteins, characterization of protein activities and structures. Proteomics techniques, such as protein microarrays, electrophoresis techniques, mass spectrometry (mass-spec), and the yeast two-hybrid system have all been widely applied in modern biomedical research.

In this section, we mainly discuss fuzzy approaches in electrophoresis and mass-spec analysis. On the other hand, fuzzy models could be broadly applied in proteomics. For example, an adaptive neuro-fuzzy inference system [Hering *et al.*, 2003; Hering *et al.*, 2004] was applied in Fourier transform infrared (FTIR) spectroscopy, which is a technique for characterization of protein secondary structure. The study shows that proteins can be accurately classified into two main classes "all alpha proteins" and "all beta proteins" (see Appendix I).

6.3.1 Electrophoresis analysis

Electrophoresis analysis can qualitatively and quantitatively investigate expression of proteins under different conditions [Gorg *et al.*, 2000]. Two dimensional (2D) electrophoresis techniques can separate extracted proteins in gel samples of a cell or tissue in two dimensions. Proteins are distributed over a rectangular area, in the form of spots, based on their molecular weights and forces under an electric field (characterized by the isoelectric point, i.e., the "pI" value). A bioinformatics problem is to map the gel spots to proteins in a species based on the 2D values. Analyses of these spots in terms of relative volume can also reveal the amount of expression of the proteins in the sample. Expressions of the same tissue under different conditions can be compared by using gels grown in those conditions.

A number of bioinformatics tools have been developed for 2D electrophoresis analysis [Marengo *et al.*, 2005]. SWISS-2DPAGE can locate the proteins on the 2D gel maps from Swiss-Prot (http://au.expasy.org/ch2d/). Melanie (http://au.expasy.org/melanie/) can analyze, annotate, and query 2D gel samples. Flicker (http://open2dprot.sourceforge.net/Flicker/) is an open-source stand-alone computer program for visually comparing 2D gel images. PDQuest (http://www.proteomeworks.bio-rad.com) is a popular commercial software package for comparing 2D gel images.

Fuzzy models have also found their applications in 2D gel analyses. As images from a 2D gel are often noisy and reproducibility may be poor, Marengo *et al.* [2003a] developed a fuzzy logic method to map the signals corresponding to the presence of proteins on the 2D maps into Gaussian membership functions. This approach allows us to assign different uncertainties for identified proteins on a 2D gel. To compare different 2D gel electrophoresis images, Kaczmarek *et al.* [2002] developed a feature-based matching technique for fuzzy alignment between gel spots across images, where the number of features in two gel images does not have to be the same. Marengo *et al.* [2003b] also proposed a fuzzy method for the comparison of different 2D maps. This approach digitized the 2D image and fuzzified the digital map. Principal component analysis and linear discriminant analysis were applied for comparing the 2D maps. This method successfully differentiated between 2D gels from healthy humans and those from non-Hodgkin lymphomas.

6.3.2 Protein identification through mass-spec

After protein separation using 2D electrophoresis or liquid chromatographic separation, protein spots are typically identified using mass-spec (MS) [Aebersold and Mann, 2003]. The data generated from mass spectrometers are often complicated and computational analyses are critical in interpreting the data for protein identification [Gras and Muller, 2001; Blueggel *et al.*, 2004]. MS gives information about molecular weight about fragments of a protein sequence. With genomic sequences widely available, the masses of protein fragments can be used to identify proteins in a biological sample through searching in a database of all possible protein fragments in the species. MS protein identification involves protein digestion using an enzyme (trypsin, pepsin, glu-C, etc.), followed by peptide mass fingerprinting (PMF) [Cottrell, 1994] or tandem mass (a.k.a. MS/MS) spectrometry analysis [Yates *et al.*, 1996]. PMF uses intact masses of digested peptides for protein identification. The MS/MS method is based on peptide fragments produced by collision-induced dissociation. While the MS/MS method is more accurate in defining peptides, it is more expensive and time-consuming than PMF. Many tools have been developed for protein identification, and the most popular ones are SEQUEST (http://fields.scripps.edu/sequest/) and Mascot (http://www.matrixscience.com/). Both of them rely on the comparison between theoretical peptides derived from the database and experimental mass spectra.

An important problem in mass-spec protein identification is to predict chemical modifications of a protein (e.g., the number of amino acids cleaved, or an amino acid change with more/less atoms). Holmes and Giddings [2004] developed a fuzzy approach to address this problem. They constructed a Web-based tool PROCLAME (http://proclame.unc.edu) using either PMF or MS/MS data. The tool explored possible combinations of chemical modifications accounting for the experimental mass with a depth-first tree search using a rule-based fuzzy logic engine. Candidates are scored and ranked. Although there may not be enough information to define the chemical modifications of a protein uniquely, the tool provides a set of candidates with ranking.

6.4 Application in Drug Design

An important application of fuzzy set theory and fuzzy logic in bioinformatics is drug design. A modern drug design often targets a protein or nucleotide (DNA/RNA) in a virus/bacterium to inhibit its function so that the virus/bacterium can be killed. Figure 6.1 shows an example of a designed drug interacting with a protein (protease) in an HIV virus. Since early 1990s, fuzzy set theory and fuzzy logic have been widely used in drug development [Hess, 1995; Sproule *et al.*, 2002].

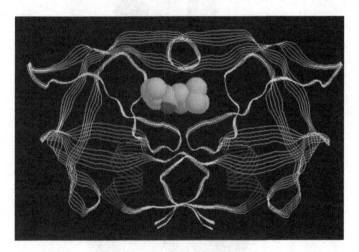

Figure 6.1 A drug molecule (shown in spheres) interacting with a HIV protein (protease, shown in ribbons).

A widely used strategy for drug design is based on the structure of potential drug chemicals (ligands) and the targeting protein. A molecule, either a ligand or a protein, can be described by its surface, as shown an example in Figure 6.2. A ligand and its target protein may form a stable complex through molecular interaction. During this event, the structure of either the ligand or the protein, including their surfaces, often changes very little. Hence, one can assume a rigid-body interaction and use the surface feature of ligands (sometimes together with the target protein) to characterize whether a ligand is a good candidate for a target protein. There are two major approaches. The first one is to perform an experimental screening using a library of ligands on a target protein and then identify the common features among the ligands that bind to the protein. These features can be used as a basis for computational studies, either to search for new ligands or to redesign the ligand for stronger interaction. The second approach is to search for the complementarity between a ligand and its target protein in terms of their molecular surfaces as a basis for drug design. This approach is referred to as docking.

Figure 6.2 Molecular surface of a protein (lysozyme). The color indicates the electric field on the protein surface.

On the other hand, how to define a surface of a molecule from the coordinates for its atoms is somewhat fuzzy [Agishtein, 1992]. The representation of a ligand (pharmacophore model) may have different degrees of "fuzziness". Renner and Schneider [2004] described such a fuzziness by a number of spheres of Gaussian-distributed feature densities. The surfaces of target proteins that can commonly bind a specific ligand may have some fuzziness in the binding pocket, especially in terms of the hydrogen-bonding pattern across the binding interface [Moodie *et al.*, 1996]. Fuzzy approaches are suitable to describe and compare molecular structures. Fuzzy logic was employed in the analysis of a database of small molecular structures [Cundari and Russo, 2001]. In particular, a fuzzy inference system was used to describe a small molecule's geometric surface that is essential for biochemical reactions, as the requirement for the geometric surface is not crisp. The study suggested a complicated interdependence among the constituent atoms in order to achieve fuzzy requirements of the geometric surface for biochemical reactions. The inference system was used for retrieving small molecules with similar structural features. Another method simplified flexible 3D chemical descriptions through clustering techniques and created "fuzzy" molecular representations called FEPOPS (feature point pharmacophores) [Jenkins *et al.*, 2004]. The representations were used for flexible 3D similarity search given one or more active ligands without a priori knowledge of bioactive features. This is similar to fuzzy protein structure comparison as discussed in Chapter 4, where a structure-alignment method was developed with a cost function containing both fuzzy assignment variables and atomic coordinates [Blankenbecler *et al.*, 2003].

Identifying features in active ligands that commonly interact with a target protein is an important subject in drug design. This is often referred to as Quantitative Structure-Activity Relationships (QSAR). One can use Comparative Molecular Surface Analysis (CoMSA) with fuzzy molecular representations as described in the last paragraph to find common surface features among active ligands for a specific target protein [reviewed in Polanski and Gieleciak, 2003]. The fuzzy features may include topographical properties, the electrostatic potential, the hydrophilicity, and the hydrogen bond density on the surface for characterization [Exner *et al.*, 2002]. Hirono *et al.* [1994a, 1994b] constructed an expert system for characterizing the pharmacokinetic properties of active ligands using the fuzzy adaptive least-squares method [Moriguchi *et al.*, 1990]. Paetz and Schneider [2005] applied a neuro-fuzzy method for classification, feature selection, and rule generation for charactering common descriptors among active ligands. Researchers also used fuzzy clustering techniques for feature selections. Lin *et al.* [2002] applied a fuzzy C-means algorithm to determine a good set of features to classify 3D convex hull descriptors computed for active HIV-1 protease inhibitors and inactive analogues. With the principal component analysis, important descriptors (feature vectors) were selected. Holliday *et al.* [2004] evaluated the use of the fuzzy C-means (FCM) clustering method for the grouping of 2D chemical structures. They demonstrated that the FCM often obtained better results than the conventional K-means method and hierarchical clustering method using Ward's distance. Berthold *et al.* [Berthold *et al.*, 2005; Wiswedel *et al.*, 2007] developed a clustering algorithm (Neighborgrams) to visualize fuzzy cluster candidates, and they applied the method to select drug candidates from chemical compounds for treating AIDS. They showed that their approach could rediscover active compounds for HIV (e.g., Azido Pyrimidines).

The fuzzy approaches discussed here are applicable not only to protein-ligand interaction, but also to RNA-ligand interaction in drug design [Renner *et al.*, 2005]. It is also worthwhile mentioning that similar fuzzy methods for molecular recognition in drug design have been applied to other areas, such as biomolecular recognition in environmental research [Exner *et al.*, 2003; Luke, 2003] and structure-camphoraceous odor relationships [Kissi *et al.*, 2004]. Uddameri and Kuchanur [2004] applied fuzzy regression methodology to study persistent organic pollutants. They investigated the sorbate-sorbent interactions using imprecise molecular descriptors based on fuzzy QSARs.

6.5 Discussion and Summary

Other than applications illustrated above, fuzzy set theory and fuzzy logic have been applied in other bioinformatics research areas, especially in biomedical literature mining and biomolecular image analysis.

Automated mining of the biomedical literature is important for retrieving key publication and information about gene function, gene-disease association, and gene-gene interactions, as it is increasingly difficult for researchers to keep current with the literature. A common goal in literature mining is to predict the relationship between two terms (words), e.g., "allergy" and "protein kinase". For this purpose, Perez-Iratxeta *et al.* [2002] developed a system XplorMed (http://www.bork.embl-heidelberg.de/xplormed/) by applying a fuzzy binary relation formalism with a standard grammatical tagger. For gene function prediction, Perez *et al.* [2004] developed a Web server (http://www.bork.embl.de/kat) based on a model of fuzzy associations to derive keywords related to a gene from literature abstracts.

Fuzzy models have been used in medical image analyses for about two decades [see reviews of Bezdek *et al.*, 1997; Toro, 2006]. In recent years, researchers started to use fuzzy approaches to study biomolecular images. Pascual *et al.* [2000] combined Kohonen's self-organizing feature maps (SOFM) and FCM in the unsupervised classification of electron microscopic images of biomolecules. The method was demonstrated to be superior to the SOFM alone for large, high-dimensional and noisy images. Mousavi *et al.* [2002] developed an iterative fuzzy algorithm for image segmentation in chromosome classification. Ferrara *et al.* [2005] developed a fuzzy method for image analysis of oligonucleotide microarrays in typing human leukocyte antigen. They used fuzzy basis functions to label image spots on microarrays.

The diverse bioinformatics applications discussed in this chapter suggest even greater potential for applying fuzzy concepts and methods in addressing a broader range of bioinformatics problems. Typically a bioinformatics problem can be solved by many methods. In some cases other techniques are better in terms of performance for accuracy or computational speed, while in others fuzzy set theory and fuzzy logic are the most suitable techniques or provide unique solutions that other methods cannot offer.

Chapter 7

Summary and Outlook

In this book, we introduced fundamentals of fuzzy set theory and fuzzy logic and their major applications in bioinformatics, including similarities in ontologies, protein structure prediction and analyses, microarray data analysis, and others. We demonstrated that fuzzy concepts and methods fit various bioinformatics problems both for representing the underlying biological mechanisms and for applying fuzzy methods as techniques for analyses and predictions. Many biological properties and concepts are fundamentally fuzzy. Fuzzy methods do not require a precise underlying model, which is often difficult to obtain in bioinformatics problems. In some of the bioinformatics applications introduced in this book, fuzzy methods are more suitable than crisp ones. In other cases, both crisp and fuzzy approaches can apply. More systematic comparisons using sizable benchmarks to compare between crisp and fuzzy methods are needed.

Although fuzzy set theory and fuzzy logic have been applied in bioinformatics, given their potential, we expect to see more and more such applications in the future. One example where fuzzy approaches can play an important role is to effectively integrate various types of data, from sequence, gene expression, protein interaction to phenotypes, each being noisy and mono-perspective, to infer biological knowledge. As more data are generated and the complexity of the data increases, the data-fusion problem brings more attention in the research community. There are statistical methods, e.g., meta-analysis, to address this issue. However, the underlying statistical models and the dependency among different data are often too difficult to handle. This may be a unique opportunity for fuzzy set theory and fuzzy logic to lead the related developments.

Many advantages for fuzzy approaches in bioinformatics have been discussed throughout this book. Nevertheless, fuzzy set theory and fuzzy logic are not "silver bullets" to solve everything. Like any other method, fuzzy models and approaches have their own limitations. When we use a fuzzy variable, it will be less accurate by nature than a crisp one. When we adopt a fuzzy description instead of a probability model, we treat the underlying mechanism more like a black box without explicit equations or probability distribution, and thus less detail is included. Fuzzy techniques sometimes ignore "high-order effects". For example, in contrast to hidden Markov models, fuzzy approaches generally do not consider correlational/transitional effects among different states. When applying fuzzy approaches, these limitations should also be considered.

Bioinformatics applications raised new challenges for fuzzy set theory. For example, there are usually a large number of free parameters in many applications. How to systematically derive suitable parameters for computational models in biological systems is non-trivial. For crisp methods, this problem has been addressed to certain extent by using methods such as orthogonal arrays [Sun et al., 1999]. The problem is not well addressed for fuzzy models while fuzzy methods often introduce more parameters to describe the fuzziness than crisp methods. Such a challenge calls for both theoretical developments in fuzzy set theory and novel integration of biological knowledge for solving the problem. Another challenge is that biologists often expect more than a fuzzy value as the overall assessment. Providing a more quantitative confidence assessment for prediction results based on the fuzzy evaluation is often important. In particular, instead of providing a fuzzy value ranging from 0 to 1, which may be hard to interpret, it would be useful to represent the value in terms of percentage of accuracy or an expectation value/p-value. Fuzzy probability theory could address this issue, but its application in bioinformatics has not been reported to our knowledge. Alternatively, in a particular bioinformatics application it may be practical through benchmarking the relationship between the fuzzy value and prediction accuracy, as done in [Bondugula and Xu, 2006]. Another challenge is the dimension of many bioinformatics problems, which is much larger than what fuzzy set theory typically addressed in the past. Many biological data often have thousands of dimensions. Applications of fuzzy approaches in these problems may require both more powerful computers and new frameworks in fuzzy set theory. Ultimately, while fuzzy set theory helps bioinformatics, biological questions will provide a driving force for new developments in fuzzy set theory itself.

Appendix I

Fundamental Biological Concepts

This Appendix introduces some fundamental biological concepts for readers without a biological background. Instead of providing a general overview, this introduction is focused only on the biological subjects discussed in the book. Our descriptions over-simplify the biological complexity. If a reader wishes to gain a more comprehensive understanding about basic biology, we recommend reading the reference materials described here and in Appendix II.

AI.1 DNA, RNA and Genome

AI.1.1 DNA (deoxyribonucleic acid)

DNA stores genetic information that controls all cellular processes through determining synthesis and regulation of proteins. DNA can reproduce itself. James Watson and Francis Crick identified its structure in 1953. The structure consists of two helical strands intertwined with each other to form the well-known double-stranded helix as shown in Figure AI.1.

A DNA sequence contains four types of units called nucleotides, i.e., adenine (A), cytosine (C), guanine (G), and thymine (T). Adenine always pairs with thymine and cytosine always pairs with guanine across the helix. A base pair means that the two nucleotides form strong hydrogen bonds, to be linked together. This complementary relationship between bases is essential in stabilizing the double helix structure. From a computer science point of view, one can imagine that one strand of a

179

DNA helix is a long "string", with the alphabet A, T, C and G. DNA passes hereditary information from one cell to another cell and from one generation to the next generation through transferring the coding information of the "string".

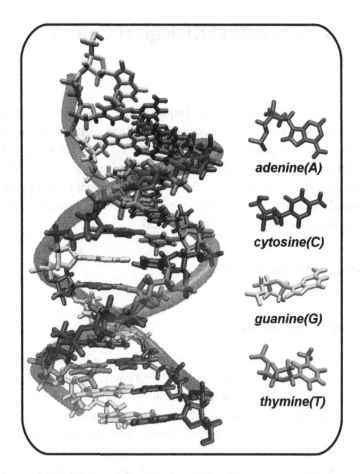

Figure AI.1 DNA structure. The left hand side shows a segment of DNA in the double-stranded helix structure. The right hand side shows the four building blocks (nucleotides) in DNA, where each line represents a chemical bond between two atoms.

AI.1.2 RNA (ribonucleic acid)

RNA has the same nucleotides as DNA except that uracil (U) replaces thymine (T). RNA is typically much shorter than DNA and often forms complicated structures with single-stranded nucleotides. There are various types of RNAs, including Messenger RNA (mRNA), Transfer RNA (tRNA), Ribosomal RNA (rRNA), Non-coding RNA (ncRNA), etc. Among them, mRNA acts as intermediate for the information transferred from the DNA to protein, which we will discuss in Section AI.3.

AI.1.3 Genome

The complete set of DNA of an organism is called its genome. All of the DNA in a species can be in a single chain or multiple chains, each of which is called a chromosome. For example, in humans, the DNA is tightly packed into 24 distinct chromosomes. The size of the genome is referred in terms of number of base pairs. The smallest genomes are viruses, often containing a few hundred thousand base pairs, while the largest genome known to date, the trumpet lily, has a genome size of about 90 billion base pairs. The size of human genome is about 3 billion base pairs. A genome is often called the blueprint of a species, as it contains the complete hereditary information of the species. The basic physical and functional unit of heredity is a gene, which is a specific sequence of nucleotide bases carrying the information required for constructing a protein and a non-protein coding product (e.g. rRNA and tRNA). The portions that code genes in a genomic sequence are referred to coding regions, while other parts of the genome are non-coding regions. In a typically genome, coding regions often represent a small percentage in a genome. For a human, only about 2% of the genome represents coding regions. Predicting the coding regions from a genomic sequence is a bioinformatics research problem, known as gene finding.

Readers can find more information about DNA, RNA, and genome in [Alberts *et al.*, 1994; Berg *et al.*, 2006].

AI.2 Protein and Its Structure

Proteins are one of the most important molecules in life. They play a variety of roles depending on their types, such as structural proteins, catalytic proteins, storage and transport proteins, regulatory proteins, immune system proteins, signaling proteins, etc. The number of protein types in a living organism often ranges from thousands to tens of thousands. For example, humans have about 30,000 protein types.

The building blocks of protein are amino acids. There are 20 types of amino acids, as described in Table AI.1. A protein is a sequence of amino acids that are linked by chemical (peptide) bonds to form a poly-peptide chain, which is referred to as protein's primary structure. Most proteins have several hundred amino acids. Short runs of these amino acids form specific configurations called secondary structures (helices, strands, sheets, coils, turns and loops). The secondary structure elements are packed together into three-dimensional (tertiary) structures. A quaternary structure may be formed when several such polypeptide chains are arranged in to a stable complex structure. The hierarchy of the four protein structural levels is illustrated in Figure AI.2.

Protein secondary structure is defined by the conformation of protein backbone. The backbone of a protein or peptide consists of repeated units with the amide nitrogen N(H), the carbon C_α, and the carbonyl carbon C($=$O). An α-helix is a major secondary structure, which is almost always right handed as found in the threads of a standard wood screw. A helix is formed when the hydrogen in the N$-$H of the nth amino acid makes a hydrogen bond with oxygen in the C$=$O of the $(n+4)^{th}$ amino acid (see Figure AI.3). This pattern of repeated bonding results in a stable α-helix. On average, there are 3.6 amino acids per turn in an α-helix. Other varieties of helices exist with slightly more and slightly less amino acids per turn.

The second major type of secondary structure is β-strand (see Figure AI.3). In a β-strand, usually 5-10 consecutive amino acids are in almost fully extended conformation. When more than one β-strand lie adjacent in space, a pleated β-sheet is formed. These are held by the hydrogen bonding between C=O groups of one strand and the N—H of the adjacent strand. If all the strands in a β-sheet run in the same biochemical direction from the start (amino-terminus) to the end (carboxy-terminus) of the protein, parallel β-sheets are formed. If alternating strands in the β-sheet run in opposite directions, anti-parallel sheets are formed.

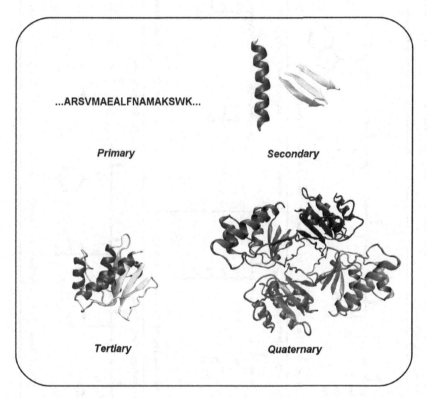

Figure AI.2 Protein structure hierarchy. The illustrated protein is an arsenate reductase from the species *Archaeoglobus* (PDB code 1Y1L).

Applications of Fuzzy Logic in Bioinformatics

Table AI.1. Twenty naturally occurring amino acids and their properties

The Twenty Natural Amino Acids Found in Proteins

A. Amino acids with hydrophobic side chains

Alanine (Ala) (A)   Isoleucine (Ile) (I)   Leucine (Leu) (L)   Methionine (Met) (M)   Phenylalanine (Phe) (F)   Tryptophan (Trp) (W)   Valine (Val) (V)

B. Amino acids with positively charged side chains

Arginine (Arg) (R)   Histidine (His) (H)   Lysine (Lys) (K)

C. Amino acids with negatively charged side chains

Aspartic acid (Asp) (D)   Glutamic acid (Glu) (E)

D. Amino acids with polar but uncharged side chains

Serine (Ser) (S)   Threonine (Thr) (T)   Asparagine (Asn) (N)   Glutamine (Gln) (Q)   Tyrosine (Tyr) (Y)

E. Special Cases

Cystine (Cys) (C)   Glycine (Gly) (G)   Proline (Pro) (P)

The core of protein tertiary structure is often formed by α-helices and β-sheets. To gain a heuristic understanding of protein tertiary structure, we compare three different representations of the same protein structure, as shown in Figure AI.4. A protein tertiary structure (including its secondary structures) is basically determined by the protein's primary sequence. Efforts to predict tertiary structure from the primary sequence are known as protein structure prediction, which has been an active research area for about three decades.

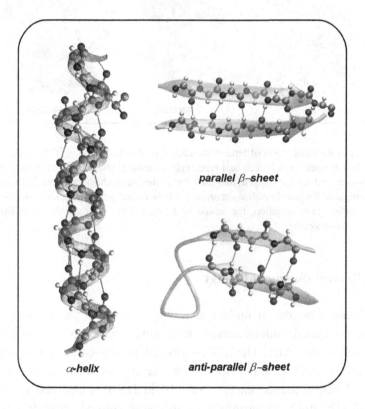

Figure AI.3 Protein secondary structures. Each ball represents an atom (light blue for carbon, dark blue for nitrogen, and red for oxygen) and each solid represents a chemical bond. An α-helix is formed when hydrogen bonds (blue dotted lines) are formed between the hydrogen atom of N-H in the nth amino acid and the oxygen of C=O of the (n+4)th amino acid. The β-sheets are held by the hydrogen bonds between the hydrogen atoms of N-H of one strand and the oxygen atoms of C=O of an adjacent strand in space.

Readers can find more information about protein and protein structures in [Branden and Tooze, 1999; Petsko and Ringe, 2004].

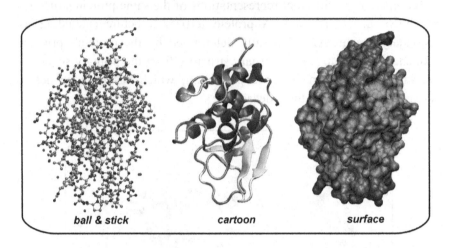

Figure AI.4 Representation of tertiary structure for protein lysozyme (PDB code 4LYZ). In the ball & stick model, each ball represents an atom (light blue for carbon, dark blue for nitrogen, and red for oxygen) and each line represents a chemical bond. In the cartoon representation, the purple ribbons represent α-helices and yellow strands show β-sheets. In the surface representation, the amino acids are colored by their distances from the center of the molecule.

AI.3 Central Dogma of Biology

The "central dogma of molecular biology" explains the mechanism by which the genetic information is transmitted from DNA to protein, as shown in Figure AI.5. The DNA gives rise to messenger RNA (mRNA) by a process called "transcription", where mRNA sequence is the same as the DNA sequence except that "T" in DNA is changed to "U" in mRNA. The mRNA transmits the information into polypeptide (protein) sequences by a process called "translation".

During the translation of an mRNA into a protein sequence, a mapping scheme exists to synthesize the protein from mRNA. This mapping scheme (Table AI.2) explains the relationship between the nucleotides in the mRNA and the amino acids in the protein. A group of three consecutive nucleotides is called a codon. As each position in a given codon can be selected from one of the four nucleotides (A,U,G,C), there are $4^3 = 64$ possible codons. Sixty one codons encode the 20 naturally occurring amino acids while the remaining 3 codons (UAA, UAG and UGA) are used as "stop" codons to let the translation machinery recognize the end of the protein. The codon "AUG", while coding amino acid methionine, is also often used as the start of a protein.

Readers can find more information about central dogma of molecular biology in [Alberts *et al.*, 1994; Lewin, 2003].

Table AI.2 Mapping scheme between three-letter codons and the twenty amino acids

1st position	*2nd position*				*3rd position*
	U	C	A	G	
U	Phe	Ser	Tyr	Cys	U
					C
	Leu		STOP		A
				Trp	G
C	Leu	Pro	His	Arg	U
					C
			Gln		A
					G
A	Ilu	Thr	Asn	Ser	U
					C
			Lys	Arg	A
	Met (START)				G
G	Val	Ala	Asp	Gly	U
					C
			Glu		A
					G

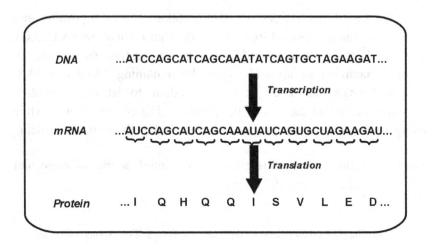

Figure AI.5 The central dogma of molecular biology. The hereditary information in the DNA is transcribed into mRNA, which in turn is translated into protein.

Appendix II

Online Resources

There is a tremendous amount of free resources related to this book on the Internet. One can easily find much useful information through keyword search at a search engine, such as Google. Here, we list and briefly describe some of the free online resources. The list given here is by no means comprehensive. Rather, we carefully selected some informative and helpful links related to the book. Please note that the cited links are active at the time of writing this book, but they may not be available or their URLs may be changed over time.

AII.1 Online Resources for Molecular Biology

The following four sites offer up-to-date information about basic molecular biology for beginners.

* DNA from the Beginning

This site features a primer of basic biological concepts through animation, image gallery, and video interviews. It has three sections, "Classical Genetics", "Molecules of Genetics", and "Genetic Organization and Control".

Web site: http://www.dnaftb.org/dnaftb/.

* MIT Biology Hypertextbook

This site provides materials for an MIT course "Introductory Biology". The materials can be searched with keywords. On-line practice problems are also available.

Web site: http://web.mit.edu/esgbio/www/.

* Wikipedia

Wikipedia is a free online encyclopedia with basic overviews. Readers can find descriptions and explanations on many basic concepts related to the book, particularly those in molecular biology, microarray technology, and fuzzy logic.

Web site: http://en.wikipedia.org.

* A Science Primer

This site gives a basic introduction to many terms related to molecular biology, microarray technology, and bioinformatics.

Web site: http://www.ncbi.nlm.nih.gov/About/primer/.

AII.2 Online Resources for Bioinformatics

There are thousands of bioinformatics sites that host biomolecular databases or prediction servers. Here, we list a few major sites that are related to the book. Readers can explore other sites from the bioinformatics portals listed in AII.2.5.

AII.2.1 Protein structure

AII.2.1.1 Protein structure database

* Protein Data Bank

Protein Data Bank (PDB) is consistent and comprehensive archive of the experimentally determined biomolecular structures. Currently, the archive contains more than 37,000 molecular structures. The website provides or links to a number of tools and resources to search, analyze and visualize biological molecules. The website also features an animated tutorial illustrating the above mentioned activities.

PDB website: http://www.rcsb.org/pdb.

AII.2.1.2. protein structure visualization

* Rasmol

Rasmol is one of oldest protein structure visualization software packages. It is a basic visualization tool with multiple options for rendering and coloring the molecules. It is now distributed as an open source program. 'Protein Explorer' is a Rasmol derivative that is more powerful and relatively easier to use.

RasMol/Protein Explorer website: www.umass.edu/microbio/rasmol.

* VMD

VMD is a powerful visualization and analysis tool for biological molecules like proteins, nucleic acids and lipid assemblies. It features a variety of rendering and coloring options. VMD also acts as a graphical interface for molecular dynamic simulations, among other external software packages.

VMD website: http://www.ks.uiuc.edu/Research/vmd.

* DeepView

DeepView/SwissPDB viewer is an application to analyze several proteins at the same time. The application can be used to study amino acid mutations, structure alignments, and hydrogen bonding. Within the application, it is possible to submit the protein for structure prediction to Swiss-Model server. Many other modeling tools are integrated into the application.

DeepView website: http://ca.expasy.org/spdbv.

AII.2.2 Microarray

AII.2.2.1 Microarray databases

* Gene Expression Omnibus (GEO)

GEO is a public repository for gene expression data. It supports retrieval and some analyses of gene expression data from any organism.

GEO website: http://www.ncbi.nlm.nih.gov/geo/.

* Stanford Microarray Database (SMD)

SMD is a database for microarray gene expression data. It provides various basic analysis capacities for the data.

SMD website: http://genome-www5.stanford.edu.

AII.2.2.2 Microarray analysis tool

* Bioconductor

Bioconductor is an open source and open development software project for the analysis of genomic data, especially micorarray data. In particular, comprehensive statistical analysis tools are available in the Bioconductor package. The development is mainly based on the R programming language.

Bioconductor website: http://www.bioconductor.org/.

* Cluster & TreeView

Cluster and TreeView form an integrated pair of programs for analyzing and visualizing microarray gene expression data. Cluster performs clustering analysis using Hierachical Clustering, K-Means Clustering, Self Organizing Maps, or PCA. The clustering results can be viewed using TreeView.

Cluster & TreeView website: http://rana.lbl.gov/EisenSoftware.htm.

* geWorkbench

geWorkbench (genomics Workbench) is a Java-based open-source platform for bioinformatics analyses. It has various components and plug-ins supporting the visualization and analysis of microarray gene expression and sequence data.

geWorkbench website: http://wiki.c2b2.columbia.edu/workbench/.

* Engene

Engene is a web-based server for the storage, analysis, and visualization of microarray gene expression data. Various clustering algorithms including K-means, HAC, fuzzy C-means, kernel C-means, SOMs, PCA, etc. are available.

Engene website: http://www.engene.cnb.uam.es.

AII.2.3 Gene ontology

* Gene Ontology (GO) Home

The site describes the results of the Gene Ontology project, which is an international collaborative effort to provide a controlled vocabulary to describe gene and gene-product attributes in any organism. The site provides annotation of biological processes and molecular functions for genes in a wide variety of species, as well as various tools for computational analyses related to GO.

GO website: http://www.geneontology.org.

AII.2.4 Online portals for bioinformatics

* The Bioinformatics Organization

The site serves the bioinformatics community through news, software access (online tools), forums, and mailing list.

Website: http://bioinformatics.org.

* Bioinformatics Links Directory

The directory provides selected links to biomolecular resources, tools and databases. It covers broad areas of bioinformatics, including sequence comparison, gene expression analysis, model organism databases, literature resources, and educational materials.

Website: http://www.bioinformatics.ubc.ca/resources/links_directory/.

* An Introduction to Bioinformatics Algorithms

This site has rich information about bioinformatics education, including a collection of course slides and problem sets, as well as links to online teaching materials in bioinformatics.

Website: http://www.bioalgorithms.info/.

AII.3 Online Resources for Fuzzy Set Theory and Fuzzy Logic

* Fuzzy Logic Tutorial

This site provides a tutorial for some basic concepts in fuzzy logic, including rule matrix, membership function, and fuzzy inference.

Website: http://www.seattlerobotics.org/encoder/mar98/fuz/flindex.html.

* Fuzzy Logic Archive

This site gives some introductory materials for fuzzy logic and links to other sites related to fuzzy logic.

Website: http://www.austinlinks.com/Fuzzy/.

Bibliography

Adleman, L.A. (2004). Molecular computation of solutions to combinatorial problems, Science, 266(11):1021–1024.

Aebersold, R. Mann, M. (2003). Mass spectrometry-based proteomics. Nature, 422:198–207.

Agishtein, M.E. (1992). Fuzzy molecular surfaces. J Biomol Struct Dyn., 9(4):759-68.

Alberts E. A., Watson, J.D. (1994). Molecular Biology of the Cell, Garland, 3rd ed.

Alon, U. (2006). An Introduction to Systems Biology, Chapman & Hall/CRC; 1st ed.

Al-Shahrour, F., Diaz-Uriarte, R., Dopazo, J. (2004). FatiGO: a web tool for finding significant associations of Gene Ontology terms with groups of genes. Bioinformatics, 20(4):578-580.

Altschul, S.F., Gish, W., Miller, W., Myers, E.W., Lipman, D.J. (1990). Basic local alignment search tool, J. Mol. Biol. 215:403-410.

Altschul, S.F., Madden, T.L., Schäffer, A.A., Zhang, J., Zhang, Z., Miller, W. and Lipman, D.J. (1997). Gapped BLAST and PSI-BLAST: a new generation of protein database search programs, Nucleic Acids Res., 25:3389-3402.

Andreasen, T., Bulskov, H. and Knappe, R. (2003). From ontology over similarity to query evaluation, 2nd CoLogNET-ElsNET Symposium - Questions and Answers: Theoretical and Applied Perspectives, Amsterdam, Holland, 200:39-50.

Andreasen, T., Bulskov, H., Knappe, R. (2003). On ontology–based querying. Proceedings of IJCAI, Acapulco, Mexico.

Anfinsen, C. (1973). Principles that govern the folding of protein chains, Science, 181:223-230.

Appel, R.D., Palagi, P.M., Walther, D., Vargas, J.R., Sanchez, J.C., Ravier, F., Pasquali, C., Hochstrasser, D.F. (1997). Melanie II -- a third-generation software package for analysis of two-dimensional electrophoresis images: I. Features and user interface. Electrophoresis, 18(15):2724-34.

Arima, C., Hanai, T., Okamoto, M. (2003). Gene expression analysis using fuzzy K-means clustering, Genome Informatics, 14:334-335.

Arredondo, T.V., Neelakanta, P.S. and De Grof, D. (2005). Fuzzy attributes of a DNA complex: development of a fuzzy inference engine for codon-"junk" codon delineation, Artif. Intell. Med., 35(1-2):87-105.

196

Arthur, G., Popescu, M., Rahmatpanah, F., Sjahputera, O., Keller, J., Shi, H., Caldwell, C. (2005). A method for simultaneous gene selection in B-cell lymphoma from methylation and expression microarrays. ISMB2005: Annual meeting of the ISCB, Detroit, Michigan.

Asyali, M.H., Alci, M. (2005). Reliability analysis of microarray data using fuzzy C-means and normal mixture modeling based classification methods. Bioinformatics, 21(5):644-649.

Atchley, W.R. and Fernandes, A.D. (2005). Sequence signatures and the probabilistic identification of proteins in the Myc-Max-Mad network, Proc. Natl. Acad. Sci. USA, 102(18):6401-6406.

Auephanwiriyakul, S., Adrian, A., Keller, J. (2002). Type 2 fuzzy set analysis in management surveys. Proceedings, Eleventh IEEE International Conference on Fuzzy Systems, Honolulu, Hawaii, pp. 1321-1325.

Auephanwiriyakul, S., Keller, J. (2002). Analysis and efficient implementation of a linguistic fuzzy C-means. IEEE Transactions on Fuzzy Systems, 10(5):563-582.

Auephanwiriyakul, S., Keller, J., Gader, P. (2002). Generalized Choquet fuzzy integral fusion, Information Fusion, 3(1):69-85.

Dasarathy, B., (ed.) (1991). Nearest Neighbor (NN) Norms: NN pattern classification techniques, IEEE Computer Society Press, Los Alamatos, CA, pp. 114-119.

Baker, P.G., Goble, C.A., Bechhofer, S., Paton, N.W., Stevens, R., Brass, A. (1999). An ontology for bioinformatics application, Bioinformatics, 15(6):510-520.

Baldi, P., Brunak, S., Frasconi, P., Pollastri, G., Soda, G. (1999). Exploiting the past and the future in protein secondary structure prediction, Bioinformatics, 15:937-946.

Baldi, P., Hatfield, W.G. (2002). DNA Microarrays and Gene Expression, Cambridge University Press.

Bandyopadhyay, S. (2005). An efficient technique for superfamily classification of amino acid sequences: feature extraction, fuzzy clustering and prototype selection, Fuzzy Sets and Systems, 152(1):5-16.

Barni, M., Cappellini, V., Mecocci, A. (1996). Comments on 'a possibilistic approach to clustering'. IEEE Trans. Fuzzy Syst., 4(3):393-396.

Belacel, N., Cuperlovic-Culf, M., Laflamme, M., Ouelette, R. (2004). Fuzzy J-Means and VNS methods for clustering genes from microarray data, Bioinformatics, 20(11):1690-1701.

Bellman, R.E., Zadeh, L.A. (1970). Decision-making in a fuzzy environment. Manage. Sci., 17(4):141-164.

Ben-Hur, A., Noble, W.S. (2005). Kernel methods for predicting protein–protein interactions, Bioinformatics, 21:i38-i46.

Berg, J.M., Tymoczko, J.L., Stryer, L. (2006) Biochemistry, Freeman, 6th ed.

Berman, H.M., Westbrook, J., Feng, Z., Gilliland, G., Bhat, T.N., Weissig, H., Shindyalov, I.N. and Bourne, P.E. (2000). The Protein Data Bank, Nucleic Acids Res., 28:235-242.

Berthold, M.R., Wiswedel, B., and Patterson, D.E. (2005). Interactive exploration of fuzzy clusters using neighborgrams, Fuzzy Sets and Systems, 149:21-37.

Bezdek, J, Keller, J., Krishnapuran, R., Pal, N. (1999). Fuzzy Models and Algorithms for Pattern Recognition and Image Processing, Kluwer Academic Publishers, Norwell, MA.

Bezdek, J.C. (1981). Pattern Recognition with Fuzzy Objective Function Algorithms, New York, NY: Plenum.

Bezdek, J.C., Hall, L.O., Clark, M.C., Goldgof, D.B., Clarke, L.P. (1997). Medical image analysis with fuzzy models. Stat Methods Med Res., 6(3):191-214.

Bezdek, J.C., Hathaway, R.J. (2002). VAT: A tool for visual assessment of (cluster) tendency, Proc. IJCNN 2002, IEEE Press, Piscataway, NJ., pp. 2225-2230.

Bezdek, J.C., Keller, J., Krishnapuram, R., Pal, N.R. (1999). Fuzzy Models and Algorithms for Pattern Recognition and Image Processing, Kluwer Academic Publishers, Massachusetts.

Black, M. (1937). Vagueness. An exercise in logical analysis. Philosophy of Science, 4(4): 427-455.

Blalock, E.M. (ed) (2003). A Beginner's Guide to Microarrays, Kluwer Academic Publishers.

Blanco, A., Pelta, D., Verdegay, J. (2002). A fuzzy valuation-based local search framework for combinatorial problems, J. Fuzzy Optim and Decision Making, 1(2):177–193.

Blankenbecler, R., Ohlsson, M., Peterson, C. and Ringner, M. (2003). Matching protein structures with fuzzy alignments, Proc. Natl. Acad. Sci. USA., 100(21):11936-11940.

Blueggel, M., Chamrad, D., Meyer, H.E.(2004). Bioinformatics in proteomics. Curr Pharm Biotechnol, 5(1):79-88.

Bolshakova1, N., Azuaje, F. and Cunningham, P. (2005). A knowledge-driven approach to cluster validity assessment, Bioinformatics, 21(10):2546–2547.

Bondugula, R., Duzlevski, O. and Xu, D. (2005). Profiles and fuzzy K-nearest neighbor algorithm for protein secondary structure prediction, Proceedings of 3rd Asia-Pacific Bioinformatics Conference, Singapore. Imperial College Press, London, pp 85-94.

Bondugula, R., Xu, D. (2007). MUPRED: A tool for bridging the gap between template based methods and sequence profile based methods for protein secondary structure prediction, Proteins, 66(3):664-70.

Bonetta, L. (2004). Bioinformatics - from genes to pathways, Nature Methods, 1:169-176.

Borkowski, L. (ed). (1970). Jan Lukasiewicz: Selected Works, Amesterdam: North-Holland.

Branden, C. and Tooze, J. (1999). Introduction to Protein Structure, 2nd ed., Garland Publishing, Inc. New York and London.

Brenner, S.E., Koehl, P., Levitt, M. (2000). The ASTRAL compendium for sequence and structure analysis. Nucleic Acids Research, 28:254-256.

Burge, C, Karlin, S. (1997). Prediction of complete gene structures in human genomic DNA, J Mol Biol, 268(1):78-94.

Cao, J., Shridhar, M., Ahmadi, M. (1995). Fusion of classifiers with fuzzy integrals. Proceedings International Conference on Document Analysis and Recognition, Montreal CA, pp. 108-111.

Caprara, A., Carr, R., Istrail, S., Lancia, G., Walenz, B. (2004). 1001 optimal PDB structure alignments: integer programming methods for finding the maximum contact map overlap, J. Comput. Biol., 11(1):27–52.

Carlacci, L., Chou, K.C., Maggiora, G.M. (1991). A heuristic approach to predicting the tertiary structure of bovine somatotropin, Biochemistry, 30(18):4389-4398.

Carr, B., Hart, W., Krasnogor, N., Burke, E., Hirst, J., Smith, J. (2002). Alignment of protein structures with a memetic evolutionary algorithm, in GECCO-2002: Proc.Genetic and Evolutionary Comput. Conf., Morgan Kaufman, Los Altos, CA, 2002.

Casasnovas, J., Rossello, F. (2005). Averaging fuzzy biopolymers. Fuzzy Sets and Systems, 152(1):139-158.

Chan, H., Dill, K. (1990) Origins of structure in globular proteins, Proc. Nat. Acad. Sci. USA 97:6388–6392.

Chandonia, J.M., Karplus, M. (1995). Neural networks for secondary structure and structural class predictions, Protein Science, 4(2):275-285.

Chang, B.C. and Halgamuge, S.K. (2002). Protein motif extraction with neuro-fuzzy optimization, Bioinformatics, 18(8):1084-90.

Chen, T., Kao, M, T., Tepel, M., Rush, J., Church, G. (2000). A dynamic programming approach to de novo peptide sequencing via tandem mass spectrometry. J Comput Biol, 8:325-337.

Chen, Y., Xu, D. (2003). Computational analyses of high-throughput protein-protein interaction data. Current Protein and Peptide Science, 4:159-181.

Cheng, H., Sen, T.Z., Kloczkowski, A., Margaritis, A., Jernigan, R.L. (2005). Prediction of protein secondary structure by mining structural fragment database. Polymer, 46:4314-4321.

Cheng,Y., Church, G.M. (2000). Biclustering of expression data. In Proceedings of the Eighth International Conference on Intelligent Systems for Molecular Biology (ISMB), pp. 93-103.

Chiang, J., Gader, P. (1997). Hybrid fuzzy-neural systems in handwritten word recognition. IEEE Trans. Fuzzy Systems, 5(4):497-510.

Cho, S.-B., Kim, J. (1995). Combining multiple neural networks by fuzzy integral for robust classification. IEEE Trans. Systems, Man, Cybernetics, 25(2):380 -384.

Choquet, G., (1953). Theory of capacities, Annales de l'Institut Fourier, 5:131–295.

Chou, P. Y., Fasman, G.D. (1974). Conformational parameters for amino acids in helical, -sheet, and random coil regions calculated from proteins, Biochemistry, 13:211-222.

Chou, P.Y. (1989). Prediction of protein structural classes from amino acid compositions, in Fasman, G.D. (ed.), Prediction of Protein Structure and the Principles of Protein Conformation, New York: Plenum, pp. 549-586.

Chu, W., Ghahramani, Z., Falciani, F., Wild, D.L. (2005). Biomarker discovery in microarray gene expression data with Gaussian processes, Bioinformatics,21(16):3385-3393.

Clauser, K.R., Baker, P.R., Burlingame, A.L. (1999). Role of accurate mass measurement (+/-10 ppm) in protein identification strategies employing MS or MS/MS and database searching. Analytical Chemistry,71:2871-2882.

Claverie, J.M. (1999). Computational methods for the identification of differential and coordinated gene expression. Human Molecular Genetics. 8:1821-1832.

Claverie, J.M., Notredame, C. (2003). Bioinformatics for Dummies, John Wiley & Sons.

Clerc, M. (2004). Discrete Particle Swarm Optimization, New Optimization Techniques in Engineering, Springer-Verlag.

Cohen, F.E., Kuntz, I.E. (1987). Prediction of the three-dimensional structure of human growth hormone. Proteins: Structure, Function, and Genetics, 2(2):162–166.

Cork, D.J., Hutch, T.B., Marland, E., Zmuda, J. (2002). Achieving congruency of phylogenetic trees generated by W-curves of genomic sequences. Ann N Y Acad Sci.,980:23-31.

Cork,D.J., Toguem, A.(2002).Using fuzzy logic to confirm the integrity of a pattern recognition algorithm for long genomic sequences: the W-curve. Ann N Y Acad Sci., 980:32-40.

Cotik,V., Romero-Záliz, R., Zwir, I. (2005). A hybrid promoter analysis methodology for prokaryotic genomes. Fuzzy Sets and Systems, 152(1):83-102.

Cottrell, J.S. (1994). Protein identification by peptide mass fingerprinting. Pept. Res. 7:115–124.

Cover, T.M. and Hart, P.E. (1967). Nearest neighbor pattern classification. IEEE Trans. on Inform. Theory., IT 13:21-27.

Cox, T.F., Cox, M.A. (2001). Multidimensional Scaling, 2nd ed. Chapman and Hall.

Cundari, T.R. and Russo, M. (2001). Database mining using soft computing techniques. An integrated neural network-fuzzy logic-genetic algorithm approach. J. Chem. Inf. Comput. Sci. 41(2):281-287.

Dancik,V., Addona, T.A., Clauser, K.R., Vath, J.E., Pevzner, P.A. (1999). De novo peptide sequencing via tandem mass spectrometry. J Comput Biol.,6(3-4):327-42.

Dave, R.N. (1991). Characterization and detection of noise in clustering, Pattern Recognition Letters, 12:657-664.

Dave, R.N., Sen, S., (2002). Robust Fuzzy Clustering of Relational Data. IEEE Trans. Fuzzy Systems, 10(6):713-727.

Dayhoff, M.O., Eck, R.V., Chang, M.A. and Sochard, M.R. (1965). Atlas of Protein Sequence and Structure Vol. 1. National Biomedical Research Foundation, Silver Spring, MD.

Deleage, G., Roux, B. (1989). Use of class prediction to improve protein secondary structure prediction, in Fasman, G.D. (ed.), Prediction of Protein Structure and the Principles of Protein Conformation. New York: Plenum, pp. 587-597.

Dembele, D., Kastner, P. (2003). Fuzzy C-means method for clustering microarray data. Bioinformatics, 19(8):973-980.

Dorigo, M., Stützle, T. (2004). Ant Colony Optimization, MIT Press.

Dougherty, E. R., Barrera, J., Brun, M., Kim, S., Cesar, R. M., Chen, Y., Bittner, M., Trent, J.M. (2002). Inference from clustering: application to gene-expression time series. Computational Biology, 9(1):105-126.

Du, P., Gong, J., Wurtele E.S., Dickerson, J.A. (2005). Modeling gene expression networks using fuzzy logic. IEEE Trans Syst Man Cybern, Part-B, 35(6):1351-1359.

Dubois, D. (ed). (2005). Special Issue on "40th Anniversary of Fuzzy Sets. Fuzzy Sets and Systems, 156(3).

Dubois, D., Prade, H. (1985). A review of fuzzy sets aggregation connectives. Inf. Sci., 36: 85-121.

Duch, W. (2005). Uncertainty of data, fuzzy membership functions, and multilayer perceptrons. IEEE Transactions on Neural Networks, 16(1):10-23.

Dunn, E., Keller, J., Marks, L., Ikerd, J., Gader, P. (1995). Extending the application of fuzzy sets to the problem of agricultural sustainability. Proceedings, ISUMA/NAFIPS'95, College Park, MD, pp. 497-502.

Eberhart, R. C., Kennedy, J. (1995). A new optimizer using particle swarm theory. Proceedings of the Sixth International Symposium on Micromachine and Human Science, Nagoya, Japan. pp. 39-43.

Eisenberg, D., Marcotte, E., McLachlan, A.D., Pellegrini, M.P. (2006). Bioinformatic challenges for the next decade(s). Trans. R. Soc. Lond. B. Biol. Sci., 361(1467):525-7.

Eng, J., McCormack, A., Yates, J. (1994). An approach to correlate tandem mass spectral data of peptides with amino acid sequences in a protein database. Journal of American Society of Mass Spectrometry, 5(11):976-989.

Ewing, B., Hillier, L., Wendl, M., and Green, P. (1998). Base calling of automated sequencer traces using phred. I. Accuracy assessment, Genome Research, 8:175-185.

Exner, T.E., Keil, M., Brickmann, J. (2002).Pattern recognition strategies for molecular surfaces. I. Pattern generation using fuzzy set theory. J Comput Chem.,23(12):1176-87.

Exner, T.E., Keil, M., Brickmann, J.(2003). New fuzzy logic strategies for bio-molecular recognition. SAR QSAR Environ Res., 14(5-6):421-31.

Fasman, G.D., (1974). Prediction of protein conformation. Biochemistry, 13(2):222-45.

Felsenstein, J. (1989). PHYLIP-Phylogeny Inference Package (Version 3.2). Cladistics, 5:164-166.

Feng, D.F., Doolittle, R.F. (1987). Progressive sequence alignment as a prerequisite to correct phylogenetic trees. J Mol Evol., 25(4):351-60.

Ferrara, G.B., Delfino,L., Masulli, F.,Rovetta,S., Sensi, R. (2005). A fuzzy approach to image analysis in HLA typing using oligonucleotide microarrays. Fuzzy Sets and Systems,152(1):37-48.

Fickett, J.W.(1982). Recognition of protein coding regions in DNA sequences, Nucleic Acids Research, 10:5303-5318.

Fitch, W. M. and Margoliash, E., (1967). Construction of phylogenetic trees. Science, 155:279 284.

Fogel, D. (2006). Evolutionary Computation: Toward a New Philosophy of Machine Intelligence, 3rd Edition, Hoboken, NJ: IEEE Press.

Fogel, D., Robinson, C. (eds) (2003). Computational Intelligence: The Experts Speak, Piscataway, NJ: IEEE Press.

Frigui, H., Nasraoui, O. (2002). Simultaneous categorization of text documents and identification of cluster dependent keywords. 2002 IEEE World Congress on Computational Intelligence. Honolulu, Hawaii.

Fukunaga, K. and Hostetler, L.D. (1975). K-nearest neighbor Bayes risk estimation. IEEE Trans. on Inform. Theory, IT 21:285-293.

Gabriele, L., Moretti, F., Pierotti, M.A., Marincola, F.M., Foà, R., Belardelli, F.M. (2006). The use of microarray technologies in clinical oncology, Journal of Translational Medicine, 4:8.

Gader, P., Frigui, H., Nelson, B., Vaillette, G., Keller, J. (2000). Fuzzy logic detection of landmines with ground penetrating radar. Signal Processing, 80(6):1069-1084.

Gader, P., Keller, J., Cai, J. (1995). A fuzzy logic system for the detection and recognition of street number fields on handwritten postal addresses. IEEE Transactions on Fuzzy Systems, 3(1):83-95.

Gader, P., Keller, J., Krishnapuram, R., Chiang, J., Mohamed, M. (1997). Neural and fuzzy methods in handwriting recognition. IEEE Computer, 30(2):79-86.

Gader, P., Keller, J., Nelson, B. (2001). Recognition technology for the detection of buried land mines. IEEE Transactions on Fuzzy Systems, 9(1):31-43.

Gader, P., Mohamed, M., Keller, J. (1996). Dynamic-programming-based handwritten word recognition using the Choquet integral as the match function. Journal of Electronic Imaging, Special Issue on Document Image Analysis, 5(1):15-24.

Gamalielsson, J., Olsson, B. (2005). GOSAP: Gene ontology based semantic alignment of biological pathways, HS-IKI-TR-05-005, Technical report, University of Skövde, Sweden.

Garwood, K.L., Taylor, C.F., Runte, K.J., Brass, A., Oliver, S.G., Paton, N.W. (2004). Pedro: a configurable data entry tool for XML. Bioinformatics, 20(15):2463-5.

Gasch, A.P., Eisen, M.B. (2002). Exploring the conditional coregulation of yeast gene expression through fuzzy k-means clustering, Genome Biology, 3(11):0059.1-0059.22.

Gath, I., Geva, A. (1989). Unsupervised optimal fuzzy clustering. IEEE Transactions on Pattern Analysis and Machine Intelligence, 11(7):773-780.

Gentleman, R.C., Carey, V.J., Bates, D.M., Bolstad, B., Dettling, M., Dudoit, S., Ellis, B., Gautier, L., Ge, Y.C., Gentry, J., Hornik, K., Hothorn, T., Huber, W., Iacus, S., Irizarry, R., Leisch, F., Li, C., Maechler, M., Rossini, A.J., Sawitzki, G., Smith, C. (2004). Bioconductor: Open software development for computational biology and bioinformatics. Genome Biology, 5(10):R80.

Geourjon, C., Deleage, G. (1994). SOPM: A self-optimized method for protein secondary structure prediction. Protein Eng, 7:157-164.

Gerstein, M., Levitt,M. (1996). Using iterative dynamic programming to obtain accurate pairwise and multiple alignments of Protein structures. Proc Int Conf Intell Syst Mol Biol 4: 59-67.

Getz, G., Levine,E., Domany, E. (2000). Coupled two-way clustering analysis of gene microarray data. Proc. Natl. Acad. Sci. USA, 97(22):12079-12084.

Giarratano, J. Riley, G., (2005). Expert Systems: Principles and Programming, 3rd Edition. Boston, PWS Publishing Company.

Gopisetty, G., Ramachandran, K., Singal, R. (2006). DNA methylation and apoptosis. Mol Immunol., 43:1729–1740.

Gorg, A., Obermaier, C., Boguth, G., Harder, A., Scheibe, B., Wildgruber, R., Weiss, W. (2000). The current state of two-dimensional electrophoresis with immobilized pH gradients. Electrophoresis, 21(6):1037-53.

Grabisch, M. (1994). Fuzzy integrals as a generalized class of order filters. Proceedings of SPIE, 2315:128-136.

Grabisch, M., Murofushi, T. and Sugeno, M. (eds) (2000). Fuzzy Measures and Fuzzy Integrals: Theory and Applications, Heidelberg: Physica-Verlag.

Grabisch, M., Murofushi, T., Sugeno, M. (1992). Fuzzy measure of fuzzy events defined by fuzzy integrals. Fuzzy Sets and Systems, 50(3):293-313.

Grabisch, M., Nicolas, J-M. (1994). Classification by fuzzy integral: Performance and tests. Fuzzy Sets and Systems 65(2&3):255-271.

Grabisch, M., Schmitt, M. (1995). Mathematical morphology, order filters, and fuzzy logic. Proceedings of the International Joint Conference of the 4th FUZZ-IEEE and the 2nd IFES, Yokohama, Japan, pp. 2103-2108.

Grabisch, M., Sugeno, M. (1992). Multi-attribute classification using fuzzy integral. Proceedings of the First IEEE Conference on Fuzzy Systems, San Diego, CA, pp. 47-54.

Gras, R., Muller, M. (2001). Computational aspects of protein identification by mass spectrometry. Current Opinion in Molecular Therapeutics, 3(6):526-532.

Green, P. (2002). Whole-genome disassembly. Proc. Natl. Acad. Sci. USA. 99:4143-4144.

Guo, J., Burger, M., Nimmrich, I., Maier, S., Becker, E., Genc, B., Duff, D., Rahmatpanah, F., Chitma-Matsiga, R., Shi, H., Berlin, K., Huang, T.H., Caldwell, C.W. (2005). Differential DNA methylation of gene promoters in small B-cell lymphomas. Am J Clin Pathol., 124(3):430-439.

Guralnik, J., Simonsick, E., Ferrucci, L., Glynn, R.J., Berkman, L.F., Blazer, D.G., Scherr, P.A., Wallace, R.B. (1994). A short physical performance battery assessing lower extremity function: association with self-reported disability and prediction of mortality and nursing home admission. J Gerontol Med Sci., 49: M85–M94.

Gustafson, E.E., Kessel, W.C. (1979). Fuzzy clustering with a fuzzy covariance matrix. In Proc. of the IEEE Conference on Decision and Control, pp.761-766.

Guthke, R., Moller, U., Hoffmann, M., Thies, F., Topfer, S. (2005). Dynamic network reconstruction from gene expression data applied to immune response during bacterial infection. Bioinformatics, 21:1626-1634.

Hagler, A.T., Honig, B. (1978). On the formation of protein tertiary structure on a computer. Proc Natl Acad Sci USA.,75(2):554-558.

Halligan, B.D., Ruotti, V., Jin, W., Laffoon, S., Twigger, S.N., Dratz, E.A.(2004). ProMoST (Protein Modification Screening Tool): a web-based tool for mapping protein modifications on two-dimensional gels. Nucleic Acids Res.,32(Web Server issue):W638-44.

Hanage, W.P., Fraser, C., Spratt, B.G. (2005). Fuzzy species among recombinogenic bacteria. BMC Biol., 7;3:6.

Hansen, P., Mladenovic, N. (2001). Variable neighborhood search. European Journal of Operational Research, 130:449-467.

Hathaway, R.J., Bezdek, J.C. (1994). NERF c-Means: Non-Euclidean relational fuzzy clustering, Pattern Recognition, 27(3): 429-437.

Hathaway, R.J., Davenport, J.W., Bezdek, J.C. (1989). Relational duals of the c-Means clustering algorithms, Pattern Recogition, 22(2):205-212.

Heger, A., Holm, L. (2003). Sensitive pattern discovery with 'fuzzy' alignments of distantly related proteins. Bioinformatics,19 Suppl 1:i130-7.

Heiden, W., Brickmann, J. (1994). Segmentation of protein surfaces using fuzzy logic. J Mol Graph., 12(2):106-15.

Heisler, L.E., Torti, D., Boutros, P.C., Watson, J., Chan, C., Winegarden, N., Takahashi, M., Yau, P., Huang, T.H., Farnham, P.J.,Jurisica,I., Woodgett, J.R., Bremner, R.,Penn, L.Z.,Der, S.D. (2005). CpG Island microarray probe sequences derived from a physical library are representative of CpG Islands annotated on the human genome. Nucleic Acids Res., 33: 2952–2961.

Hering, J.A., Innocent, P.R., Haris, P.I. (2003). Neuro-fuzzy structural classification of proteins for improved protein secondary structure prediction. Proteomics, 3(8):1464-75.

Hering, J.A., Innocent, P.R., Haris, P.I. (2004). Beyond average protein secondary structure content prediction using FTIR spectroscopy. Applied Bioinformatics, 3(1): 9-20.

Hess, J. (1995). Fuzzy logic in drug delivery. Med Device Technol., 6(10):23-27.

Hinds, D.A., Stuve, L.L., Nilsen, G.B., Halperin, E., Eskin, E., Ballinger, D.G., Frazer, K.A., Cox, D.R. (2005). Whole-genome patterns of common DNA variation in three human populations. Science, 307(5712):1072-1079.

Hirono, S., Nakagome, I., Hirano, H., Matsushita, Y., Yoshii, F., Moriguchi, I. (1994).Non-congeneric structure-pharmacokinetic property correlation studies using fuzzy adaptive least-squares: oral bioavailability. Biol Pharm Bull., 17(2):306-9.

Hirono, S., Nakagome, I., Hirano, H., Yoshii, F., Moriguchi, I. (1994). Non-congeneric structure-pharmacokinetic property correlation studies using fuzzy adaptive least-squares: volume of distribution. Biol Pharm Bull., 17(5):686-90.

Hobohm, U, Sander, C. (1994). Enlarged representative set of protein structures. Protein Science, 3:522-524.

Holley, L. H. and Karplus, M. (1989). Protein secondary structure prediction with a neural network. Proc. Natl. Acad. Sci. USA, 86:152-156.

Holliday, J.D., Rodgers, S.L., Willett, P., Chen, M.Y., Mahfouf, M., Lawson, K., Mullier, G. (2004). Clustering files of chemical structures using the fuzzy k-means clustering method. J Chem Inf Comput Sci., 44(3):894-902.

Holm, L., Ouzounis, C., Sander, C., Tuparev, G., Vriend, G. (1992). A database of protein structure families with common folding motifs. Protein Science, 1:1691-1698.

Holm, L. Sander, C. (1993). Protein Structure Comparison by Alignment of Distance Matrices. J. Mol. Biol., 233(1):123-138.

Holm, L., Sander, C. (1996). Mapping the protein universe. Science, 273:595-603.

Holmes, M.R., Giddings, M.C. (2004). Prediction of posttranslational modifications using intact-protein mass spectrometric data. Anal. Chem., 6(2):276-282.

Hoogland, C., Mostaguir, K., Sanchez, J.C., Hochstrasser, D.F., Appel, R.D. (2004). SWISS-2DPAGE, ten years later. Proteomics, 4(8):2352-2356.

Horrocks, I., Patel-Schneider, P.F. (1998). FaCT and DLP. In H. de Swart, editor, Automated Reasoning with Analytic Tableaux and Related Methods: International Conference Tableaux'98, in Lecture Notes in Artificial Intelligence, 1397:27-30.

Huang, T.H., Perry, M.R., Laux, D.E. (1999). Methylation profiling of CpG islands in human breast cancer cells. Hum. Mol. Genet., 8:459–470.

Huang, Y. and Li, Y. (2004). Prediction of protein subcellular locations using fuzzy k-NN method. Bioinformatics, 1-20(1):21-8.

Ignizio, J. (1991). Introduction to Expert Systems. McGraw-Hill.

Jackson, Peter. (1998). Introduction to Expert Systems. Third Edition. Addison Wesley.

Jacob, E., Sasikumar, R. and Nair, K.N. (2005). A fuzzy guided genetic algorithm for operon prediction. Bioinformatics, 21(8):1403-7.

Jansen, R. and Gerstein, M. (2004). Analyzing protein function on a genomic scale: the importance of gold-standard positives and negatives for network prediction. Curr. Opin. Microbiol., 7:535-45.

Jenkins, J.L., Glick, M., Davies, J.W. (2004). A 3D similarity method for scaffold hopping from known drugs or natural ligands to new chemotypes. J. Med. Chem., 47(25):6144-6159.

Jiang, F. (2003). Prediction of protein secondary structure with a reliability score estimated by local sequence clustering. Protein Eng, 16:651-657.

Jiang, J.J., Conrath, D.W. (1997). Semantic similarity based on corpus statistics and lexical ontology. Proc. of Int. Conf. Research on Comp. Linguistics X, Taiwan.

Jiang, T., Xu, Y., Zhang, M.Q. (2002) Current Topics in Computational Molecular Biology, MIT Press/Tsinghua Press.

Jones, D.T. (1999). Protein secondary structure prediction based on position-specific scoring matrices. J. Mol. Biol., 292:195-202.

Jones, N.C., Pevzner, P.A. (2004). An Introduction to Bioinformatics Algorithms, MIT Press.

Joslyn, C.A., Mniszewski, S.M., Fulmer, A. and Heaton, A. (2004). The gene ontology categorizer, Bioinformatics, 20(1):69–77.

Kabsch, W., Sander, C. (1983). Dictionary of protein secondary structure: pattern recognition of hydrogen-bonded and geometrical features. Biopolymers, 22:2577-2637.

Kacprzyk, J., Ziolkowski, A. (1986). Retrieval from data bases using queries with linguistic quantifiers, in Fuzzy Logic in Knowledge Engineering, H. Prade and C. V. Negoita, eds. Cologne, Germany: Verlag TUV Rheinland, pp. 46–57.

Kaczmarek, K., Walczak, B., de Jong, S., Vandeginste, B.G. (2002). Feature based fuzzy matching of 2D gel electrophoresis images. J Chem Inf Comput Sci., 42(6):1431-42.

Kanehisa, M., Bork P. (2003). Bioinformatics in the post-sequence era. Nat Genet, 33 Suppl:305-310.

Kanehisa, M., Goto, S. (2000). KEGG: Kyoto Encyclopedia of Genes and Genomes. Nucleic Acids Research, 28:27–30.

Karplus, K., Barrett, C., Hughey, R. (1998). Hidden Markov models for detecting remote protein homologies. Bioinformatics, 14:846-856.

Karplus, M., Weaver, D.L. (1976). Protein-folding dynamics. Nature, 260(5550):404-6.

Keller, J. (1997). Fuzzy set theory in computer vision: A prospectus. Fuzzy Sets and Systems. 90(2):177-182.

Keller, J. Gader, P., Hocaoglu, A. K. (2000). Fuzzy integrals in image processing and recognition, in Fuzzy Measures and Integrals: Theory and Applications, M. Grabisch, T. Murofushi, and M. Sugeno (eds.). Springer-Verlag, 2000, pp. 435-466.

Keller, J., Bezdek, J., Popescu, M., Pal, N., Mitchell, J., Huband, J. (2005). OWA operators for gene product similarity, clustering, and knowledge discovery. Proceedings, NAFIPS 2005, Ann Arbor, MI, pp. 233-234.

Keller, J., Bezdek, J., Popescu, M., Pal, N., Mitchell, J., Huband, J. (2006). gene ontology similarity measures based on linear order statistics. International Journal on Uncertainty, Fuzziness and Knowledge-Based Systems, 14(6):639-661.

Keller, J., Carpenter, C. (1988). Image segmentation in the presence of uncertainty. proceedings NAFIPS-88, San Francisco, CA, pp. 136-140.

Keller, J., Carpenter, C. (1990). Image segmentation in the presence of uncertainty. International Journal of Intelligent Systems, 5(2):193-208.

Keller, J., Downey, T. (1988). Fuzzy segmentation using fractal features. Proceedings, SPIE Symposium on Intelligent Robots and Computer Vision, Cambridge, MA, pp. 369-376.

Keller, J., Gader, P. (1995). Fuzzy logic and the principle of least commitment in computer vision. Proceedings, IEEE Conference on Systems, Man, and Cybernetics, Vancouver, BC, pp. 4621-4625.

Keller, J., Gader, P., Caldwell, C.W. (1995). The principle of least commitment in the analysis of chromosome images. Proceedings, SPIE Symposium on OE/Aerospace Sensing and Dual Use Photonics, Orlando, FL, pp. 178-186.

Keller, J., Gader, P., Krishnapuram, R., Wang, X., Hocaoglu, A. K., Frigui, H., Moore, J. (1998). A fuzzy logic automatic target recognition system for ladar range images. Proceedings, Seventh IEEE International Conference on Fuzzy Systems, Anchorage, Alaska, pp. 71-76.

Keller, J., Gader, P., Sjahputera, O., Caldwell, C.W. Huang, H-M. (1995). A fuzzy logic rule-based system for chromosome recognition. Proceedings, Eighth IEEE Symposium on Computer-Based Medical Systems, Lubbock, TX, pp. 125-132.

Keller, J., Gader, P., Tahani, H., Chiang, J-H., Mohamed, M. (1994). Advances in fuzzy integration for pattern recognition. Fuzzy Sets and Systems, Special Issue on Pattern Recognition, 65(2&3):273-283.

Keller, J., Gray, M., Givens, J. (1985). A fuzzy K nearest neighbor algorithm. IEEE Transactions on Systems, Man, and Cybernetics, 15(4):580-585.

Keller, J., Hayashi, Y., Chen, Z. (1994). Additive hybrid networks for fuzzy logic. Fuzzy Sets and Systems, 66(3):307-313.

Keller, J., Hobson, G., Wootton, J., Nafarieh, A., Luetkemeyer, K. (1987). Fuzzy confidence measures in midlevel vision. IEEE Transactions on Systems, Man, and Cybernetics, 17(4):676-683.

Keller, J., Hunt, D. (1985). Incorporating fuzzy membership functions into the perceptron algorithm. IEEE transactions on Pattern Analysis Machine Intelligence, 7(6): 693-699.

Keller, J., Krishnapuram, R. (1992). Fuzzy set methods in computer vision, An Introduction to Fuzzy Logic Applications in Intelligent Systems, R. Yager and L. Zadeh (eds)., Kluwer Academic Press, pp. 121-146.

Keller, J., Krishnapuram, R. (1994). Fuzzy decision models in computer vision, in Fuzzy Sets, Neural Networks, and Soft Computing, R. Yager and L. Zadeh (eds.), Van Nostrand, pp. 213-232.

Keller, J., Krishnapuram, R., Chen, Z., Nasraoui, O. (1994). Fuzzy additive hybrid operators for network-based decision making. International Journal of Intelligent Systems, 9(11):1001-1024.

Keller, J., Krishnapuram, R., Gader, P., and Choi, Y-S. (1996). Fuzzy Rule-based models in computer vision, in Fuzzy Modelling: Paradigms and Practice, W. Pedrycz (ed), Kluwer Academic Publishers, 1996, pp. 353-371.

Keller, J., Krishnapuram, R., Rhee, F. (1992). Evidence aggregation networks for fuzzy logic inference. IEEE Transactions on Neural Networks, 3(5):761-769.

Keller, J., Matsakis, P. (1999). Aspects of high level computer vision using fuzzy sets. Proceedings, Eighth IEEE International Conference on Fuzzy Systems, Seoul, Korea, pp. 847-852.

Keller, J., Osborn, J. (1996). Training the fuzzy integral. international journal of approximate reasoning, 15(1):1-24.

Keller, J., Popescu, M., Mitchell, J. (2004). Taxonomy-based soft similarity measures in bioinformatics. Proceedings, 13th IEEE International Conference on Fuzzy Systems, Budapest, Hungary, pp. 23-30.

Keller, J., Qiu, H., Tahani, J. (1986). The fuzzy integral in image segmentation. Proceedings, NAFIPS-86, New Orleans, pp. 324-338.

Keller, J., Qiu, H. (1988). Fuzzy set methods in pattern recognition. Pattern Recognition, Lecture Notes in Computer Science, Vol 301, J. Kittler, ed., Springer-Verlag, Berlin, pp.173-182.

Keller, J., Tahani, H. (1992). Backpropagation neural networks for fuzzy logic. Information Sciences, 62(3):205-221.

Keller, J., Tahani, H. (1992). Implementation of conjunctive and disjunctive fuzzy logic rules with neural networks, International Journal of Approximate Reasoning, Special issue on "Fuzzy Logic and Neural Networks for Control", 6(2):221-240.

Keller, J., Wang, X. (2000). A fuzzy rule-based approach for scene description involving spatial relationships. Computer Vision and Image Understanding, 80(1):21-41.

Keller, J., Yager R., Tahani, H. (1992). neural network implementation of fuzzy logic. Fuzzy Sets and Systems, 45(1):1-12.

Keller, J.M., Gray, M.R. and Givens, Jr, J. A. (1985). A fuzzy K-nearest neighbor algorithm. IEEE Trans. On SMC, 15(4):580.

Kennedy, J., Eberhart, R. C. (1995). Particle swarm optimization. Proceedings of IEEE International Conference on Neural Networks, Piscataway, NJ. pp. 1942-1948.

Khan, S., Situ, G., Decker, K., Schmidt, CJ. (2003). GoFigure: automated Gene Ontology annotation. Bioinformatics. 19(18):2484-2485.

Khatri, P., Bhavsar, P., Bawa, G., Draghici, S. (2004). Onto-Tools: an ensemble of web-accessible, ontology-based tools for the functional design and interpretation of high-throughput gene expression experiments. Nucleic Acids Res. 32:W449-456.

Khatri, P., Draghici, S., Ostermeier, G.C. and Krawetz, S.A. (2002). Profiling gene expression using Onto-Express, Genomics, 79(2):266-270.

King, R. D., Saqi, M., Sayle., R., Sternberg, M.J. (1997). DSC: public domain protein secondary structure prediction. Comput Appl Biosci, (13):473-474.

King, R. D., Sternberg, M. J. (1996). Identification and application of the concepts important for accurate and reliable protein secondary structure prediction, Protein Science, 5:2298-2310.

Kissi, M., Ramdani, M., Tollabi, M., Zakarya, D. (2004). Determination of fuzzy logic membership functions using genetic algorithms: application to structure-odor modeling. J Mol Model, 10(5-6):335-41.

Klir, G., Yuan, B. (1995). Fuzzy Sets and Fuzzy Logic: Theory and Applications, New Jersey: Prentice Hall.

Kriete, A., Eils, R. (eds.) (2005). Computational Systems Biology, Elsevier.

Krishnapuram, R., Joshi, A., Nasraoui, O., Yi, L. (2001). Low-complexity fuzzy relational clustering algorithms for web mining. IEEE Trans. Fuzzy Systems, 9(8):595-607.

Krishnapuram, R., Keller, J. (1993). A Possibilistic Approach to Clustering, IEEE Transactions on Fuzzy Systems, 1(2):98-110.

Krishnapuram, R., Keller, J. (1994). Fuzzy and Possibilistic Clustering Methods for Computer Vision, in Neural and Fuzzy Systems, S. Mitra, M. Gupta, and N. Kraske (eds.), SPIE Press, pp. 133-159.

Krishnapuram, R., Keller, J. (1996). The possibilistic C-means algorithm: Insights and recommendations. IEEE Transactions on Fuzzy Systems, 4(3):385-393.

Krishnapuram, R., Lee, J. (1992a). Fuzzy-set-based hierarchical networks for information fusion in computer vision. Neural Netw., 5(2):335-350.

Krishnapuram, R., Lee, J. (1992b). Fuzzy-set-based hierarchical networks for information fusion for Decision Making. Fuzzy Sets and Systems, 46(1):11-27.

Kummamuru, K., Dhawale, A., Krishnapuram, R. (2003). Fuzzy co-clustering of documents and keywords. IEEE International Conf. on Fuzzy Systems, 2:772-777.

Kuntz, I., Blaney, J., Oatley, S., Langridge, R., Ferrin, T. (1982). A geometric approach to macromolecule-ligand interactions. Journal of Molecular Biology, 161:269-288.

Laskowski, R.A. Structural quality assurance, in Structural Bioinformatics, Wiley-Liss, New York, 2003.

Lau, R.Y.K. (2007). Ontology extraction for business knowledge management. Proceedings of the 2007 International Conference on Business and Information (BAI'07), July 11-13, 2007, Tokyo, Japan.

Lazzeroni, L.C, Owen, A. (2002). Plaid models for gene expression data. Statist Sinica, 12(1):61–86.

Lee, M.L.T. (2004). Analysis of Microarray Gene Expression Data, Kluwer Academic Publishers, Boston.

Lee, B, Richards, F.M. (1971). The interpretation of protein structures: estimation of static accessibility. J Mol Biol., 55(3):379-400.

Lee, C.S., Jian, Z.W., Huang L.K. (2005). A fuzzy ontology and its application to news summarization. IEEE Trans. on Systems, Man and Cybernetics, Part B, 35(5):859-880.

Lee, J. (2006). Measures for the assessment of fuzzy predictions of protein secondary structure. Proteins, 65(2):453-462.

Lemkin, P.F. (1997). Comparing two-dimensional gels across the Internet. Electrophoresis, 18:461-470.

Leng, Q. and Bentwich, Z. (2002). Beyond self and nonself: fuzzy recognition of the immune system. Scand. J. Immunol., 56:224-232.

Lesk, A.M. (2005). Introduction to Bioinformatics, Oxford University Press, 2nd ed.

Levitt, M. and Warshel, A. (1975). Computer simulation of protein folding. Nature, 253:694-698.

Levy, S.E. (2003). Microarray analysis in drug discovery: an uplifting view of depression, Sci STKE., 2003(206):46.

Lewin, B. (2003). Genes VIII, Prentice Hall.

Liang, S., Samanta, M.P., Biegel, B.A. (2004). cWINNOWER algorithm for finding fuzzy dna motifs. J Bioinform Comput Biol., 2(1):47-60.

Lin D. (1998). An information-theoretic definition of similarity, In Proc. 15th International Conf. on Machine Learning, San Francisco, CA. pp. 296-304.

Lin, T.H., Wang, G.M, Hsu, Y.H. (2002). Classification of some active HIV-1 protease inhibitors and their inactive analogues using some uncorrelated three-dimensional molecular descriptors and a fuzzy c-means algorithm. J Chem Inf Comput Sci., 42(6):1490-504.

Lipman, D.J., Pearson, W.R. (1985). Rapid and sensitive protein similarity searches. Science, 227(4693):1435-41.

Lord P.W., Stevens R., Brass A., Goble C.A. (2003). Investigating semantic similarity measures across the Gene Ontology: the relationship between sequence and annotation. Bioinformatics, 19(10):1275-1283.

Luke, B.T. (2003). Fuzzy structure-activity relationships. SAR QSAR Environ Res., 14(1):41-57.

Luscombe, N.M.,Greenbaum,D., Gerstein, M. (2001). What is bioinformatics? A proposed definition and overview of the field. Methods Inf Med, 40: 346-58.

Ma, B., Zhang, K., Lajoie, G., Doherty-Kirby, C., Hendrie, C., Liang, C., Li, M. (2003). Peaks:powerful software for peptide de novo sequencing by tandem mass spectrometry. Rapid Communication in Mass Spectrometry, 17(20):2337-2342.

Madeira, S. C. and Oliveira, A. L. (2004). Biclustering algorithms for biological data analysis: a survey. IEEE/ACM Transactions on Computational Biology and Bioinformatics, 1(1):24:45.

Mamdani, E. H. (1977). Applications of Fuzzy logic to approximate reasoning using linguistic systems. IEEE Transactions on Systems, Man, and Cybernetics, 26(12):1182-1191.

Mamdani, E. H., Assilian, S. (1999). An experiment in linguistic synthesis with a fuzzy logic controller. Int. J. Hum.-Comput. Stud. 51(2):135-147.

Mamdani, E. H., Assilian, S. (1975). An experiment in linguistic synthesis with a fuzzy logic controller", Int. Journal of Man-Machine Studies, 7:1-13.

Manning, C.D., Schutze, H. (2001). Foundations of Statistical Natural Language Processing, MIT Press.

Marengo, E., Robotti, E., Antonucci, F., Cecconi, D., Campostrini, N., Righetti, P.G. (2005). Numerical approaches for quantitative analysis of two-dimensional maps: a review of commercial software and home-made systems. Proteomics, 5(3):654-66.

Marengo, E., Robotti, E., Gianotti, V., Righetti, P.G. (2003). A new approach to the statistical treatment of 2D-maps in proteomics using fuzzy logic. Ann Chim., 93(1-2):105-116.

Marengo, E., Robotti, E., Righetti, P.G., Antonucci, F. (2003). New approach based on fuzzy logic and principal component analysis for the classification of two-dimensional maps in health and disease. Application to lymphomas. J Chromatogr A., 1004(1-2):13-28.

Margulies, M., Eghold, M., Altman, W.E., Attiya, S., Bader, J.S., Bemben, L.A., Berka, J., Braverman, M.S., Chen, Y., Chen, Z., Dewell, S.B., Du, L., Fierro, J.M., Gomes, X.V., Godwin, B.C., He, W., Helgesen, S., Ho, C.H., Irzyk, G.P., Jando, S.C., Alenquer, M.L.I., Jarvie, T.P, Jirage, K.B., Kim, J.B., Knight, J.R., Lanza, J.R., Leamon, J.H., Lefkowitz, S.H., Lei, M., Li, J., Lohman, K.L., Lu ,H., Makhijani, V.B., McDade, K.E., McKenna, M.P., Myers, E.W., Nickerson,E., Nobile, J.R., Plant, R., Puc, B.P., Ronan, M.T., Roth, G.T., Sarkis, G.J., Simons, J.F., Simpson, J.W., Srinivasan, M., Tartaro, K.R., Tomasz, A., Vogt, K.A., Volkmer, G.A., Wang, S.H., Wang, Y., Weiner, M.P., Yu, P.,Begley, R.F., Rothberg, J.M. (2005). Genome sequencing in microfabricated high-density picolitre reactors. Nature, 437(7057):326-7.

Marks, L., Dunn, E., Keller, J. (1995). Multiple criteria decision making (MCDM) using fuzzy logic: an innovative approach to sustainable agriculture. Proceedings, ISUMA/NAFIPS'95, College Park, MD, pp. 503-508.

Marr. D. (1982). Vision, San Francisco, W. H. Freeman and Company.

Martin, D.M., Berriman, M., Barton, G.J. (2004). GOtcha: a new method for prediction of protein function assessed by the annotation of seven genomes. BMC Bioinformatics, 20(5):178.

MathWorks (1995). Fuzzy Logic Toolbox for use with MATLAB – User's Guide. The Math Works, Massachusetts, US.

Mathews, D.H., Sabina, J., Zucker, M., and Turner, H. (1999). Expanded sequence dependence of thermodynamic parameters provides robust prediction of RNA secondary structure, J. Mol. Biol., 288:911-940.

Matthews, B.W. (1975). Comparison of the predicted and observed secondary structure of T4 phage lysozyme. Biochim.Biophysics.Acta.405:402-451.

Meiler, J. and Baker, D. (2003). Coupled prediction of protein secondary and tertiary structure. Proc. Natl. Acad. Sci., 100:12105-12110.

Mendez-Vazquez, A., Gader, P., Keller, J., Chamberlin, K., (2007, in press). Minimum classification error training for Choquet integrals with applications to landmine detections, IEEE Transactions on Fuzzy Systems.

Mitra, S., Hayashi, Y. (2006). Bioinformatics with soft computing. IEEE Trans Systems, Man, and Cybernetics, Part C, 36:616-635.

Mocz, G. (1995). Fuzzy cluster analysis of simple physicochemical properties of amino acids for recognizing secondary structure in proteins. Protein Sci., 4(6):1178-87.

Model, F., Adorjan, P., Olek, A., Piepenbrock, C. (2001). Feature selection for DNA methylation based cancer classification. Bioinformatics, 17 Suppl 1:S157-64.

Mogre, A., McLaren, R., Keller, J., Krishnapuram, R. (1994). Uncertainty management in rule based systems with applications to image analysis. IEEE Transactions, Systems, Man, and Cybernetics, 24(3): 470-481.

Moodie, S.L., Mitchell, J.B., Thornton, J.M. (1996). Protein recognition of adenylate: an example of a fuzzy recognition template. J Mol Biol., 263(3):486-500.

Moriguchi, I., Hirono, S., Liu, Q.A., Matsushita, Y., Nakagawa, T. (1990). Fuzzy adaptive least squares and its use in quantitative structure-activity relationships. Chem Pharm Bull., 38(12):3373-9.

Mousavi, P., Ward, R.K., Fels, S.S., Sameti, M., Lansdorp, P.M. (2002). Feature analysis and centromere segmentation of human chromosome images using an iterative fuzzy algorithm. IEEE Trans Biomed Eng., 49(4):363-71.

Murofushi, T. Sugeno, M. (1991). A theory of fuzzy measures: representations, the choquet integral, and null sets. J. Math Analysis and Applications 159:532-549.

Murofushi, T., Sugeno, M. (1991). A Theory of Fuzzy Measures: Representations, the Choquet integral and null sets. J. Math. Anal. Appl. 159:532-549.

Murzin A. G., Brenner S. E., Hubbard T., Chothia C. (1995). SCOP: a structural classification of proteins database for the investigation of sequences and structures. J. Mol. Biol., 247:536-540.

Myers, E.W. (1995). Toward simplifying and accurately formulating fragment assembly. J Comput Biol., 2:275-90.

Myllyharju, J., Kivirikko, K.I. (2004). Collagens, modifying enzymes and their mutations in humans, flies and worms. Trends in Genetics, 20(1): 33-43.

Nafarieh, A., Keller, J. (1987). Incorporating Confidence Measures into Fuzzy Classifiers. Proceedings, NAFIPS-87, Purdue University, pp. 81-94.

Nafarieh, A., Keller, J. (1991). A Fuzzy Logic Rule-Based Automatic Target Recognizer. International Journal of Intelligent Systems, 6(3): 295-312.

Nafarieh, A., Keller, J. (1991). A New Approach to Inference in Approximate Reasoning. Fuzzy Sets and Systems, 41(1):17-37.

Needleman, S.B., Wunsch, C.D. (1970). A general method applicable to the search for similarities in the amino acid sequence of two proteins. J. Mol. Biol., 48:443-453.

Neumann, E. (2005). A life science Semantic Web: are we there yet? Sci STKE 2005: pe22

Nicholls, A., Sharp, K., Honig, B. (1991). Protein folding and association: Insights from the interfacial and thermodynamic properties of hydrocarbons. Proteins: Structure, Function and Genetics, 11(4):281-286.

Nielsen, H., Engelbrecht, J., Brunak, S., von Heijne, G. (1997). Identification of prokaryotic and eukaryotic signal peptides and prediction of their cleavage sites. Protein Eng., 10(1):1-6.

Nieto, J.J., Torres, A., Georgiou, D.N., Karakasidis, T.E. (2006). Fuzzy polynucleotide spaces and metrics. Bull Math Biol., 68(3):703-725.

Nieto, J.J., Torres, A., Vazquez-Trasande, M.M. (2003). A metric space to study differences between polynucleotides. Appl. Math. Lett. 27:1289-1294

Noble, W.S. (2004). Support Vector machines applications in computational biology, in Kernel Methods in Computational Biology, B. Schoelkopf, K. Tsuda, J.P.Vert (eds), MIT Press, Cambridge, MA, pp. 71-92.

Oh, C.H., Honda, K., Ichihashi, H. (2001). Fuzzy clustering for categorical multivariate data, Proc. of the Joint 9th IFSA World Congress and 20th NAFIPS International Conference, pp.2154-2159.

Orengo, C.A., Michie, A.D., Jones, S., Jones, D.T., Swindells, M.B., Thornton, J.M. (1997). CATH -- A hierarchic classification of protein domain structures. Structure, 5(8):1093-1108.

Ouzounis, C.A., Valencia, A. (2003). Early bioinformatics: the birth of a discipline-a personal view. Bioinformatics, 19(17):2176-90.

Paetz, J., Schneider, G. (2005). A neuro-fuzzy approach to virtual screening in molecular bioinformatics. Fuzzy Sets and Systems, 152(2005):67-82.

Pal, K., Pal, N., Keller, J. (1998). Some Neural Net Realizations of Fuzzy Reasoning. International Journal of Intelligent Systems, 13(9):859-886.

Pal, K., Pal, N., Keller, J., Bezdek, J. (2005). Relational mountain (density) clustering method and web log analysis. International Journal of Intelligent Systems, 20(3):375-392.

Pal, N., Keller, J., Popescu, M., Bezdek, J., Mitchell, J., Huband, J. (2005). Gene ontology-based knowledge discovery through fuzzy cluster analysis. Journal of Neural, Parallel and Scientific Computing, 13(3-4):337-361.

Pal, N., Pal, K., Keller, J., Bezdek, J. (2004). A new hybrid c-means clustering model, Proceedings, 13th IEEE International Conference on Fuzzy Systems, Budapest, Hungary, pp. 179-184.

Pal, N., Pal, K., Keller, J., Bezdek, J. (2005). A possibilistic fuzzy c-means clustering algorithm. IEEE Transactions on Fuzzy Systems, 13(4):517 - 530.

Pal, N.R., Keller, J.M., Popescu, M., Bezdek, J.C., Mitchell, J.A. and Huband, J. (2005). Gene ontology-based knowledge discovery through fuzzy cluster analysis, Neural, Parallel and Scientific Computation, 13(3-4):337-361.

Palsson, B.O. (2006). Systems Biology-Properties of Reconstructed Networks, Cambridge University Press.

Parekh, G., Keller, J. (2007). Learning the fuzzy connectives of a multilayer network using particle swarm optimization. Proceedings, 2007 IEEE Symposium on Foundations of Computational Intelligence (FOCI 2007), Honolulu, Hawaii, pp. 591-596.

Pappin, D.J.C., Hojrup, P., Bleasby, A.J. (1993). Rapid identification of proteins by peptide-mass fingerprinting. Current Biology, 3(6):327-332.

Park, J-S., Keller, J. (1997). Fuzzy patch label relaxation in bone marrow cell segmentation. Proceedings, IEEE International Conference on Systems, Man, and Cybernetics, Orlando, FL, pp. 1133-1138.

Pascual, A., Barcena, M., Merelo, J.J. and Carazo, J.M. (2000). Mapping and fuzzy classification of macromolecular images using self-organizing neural networks. Ultramicroscopy, 84(1-2):85-99.

Passino, K., Yurkovich, S. (1998). Fuzzy Control, Menlo Park, Addison Wesley.

Paszko, C., Turner, E. (2001) Laboratory Information Management Systems, 2nd ed, CRC.

Pedraza, J.M. and van Oudenaarden, A. (2005). Noise propagation in gene networks. Science, 307:1965-1969.

Pedrycz, W., Gomide, F. (1998). An Introduction to Fuzzy Sets: Analysis and Design, Cambridge, MA: MIT Press.

Pelta, D.A, Krasnogor, N., Bousoño-Calzón, C., Verdegay, J.L., Hirst, J.D., Burke, E.K. (2005). A fuzzy sets based generalization of contact maps for the overlap of protein structures. Fuzzy Sets and Systems, 152(1):103-123.

Perez, A.J., Perez-Iratxeta, C., Bork, P., Thode, G., Andrade, M.A. (2004). Gene annotation from scientific literature using mappings between keyword systems. Bioinformatics, 20(13):2084-2091.

Perez-Iratxeta, C., Keer, H.S., Bork, P., Andrade, M.A. (2002). Computing fuzzy associations for the analysis of biological literature. Biotechniques, 32(6):1380-2, 1384-1385.

Petsko, G. A. and Ringe, D. (2004). Protein Structure and Function, New Science Press Ltd, London.

Pevzner, P.A., Dancík,V., Tang,C. (2000). Mutation tolerant protein identification by mass spectrometry, J. Comput. Biol., 6(2000):777-787.

Pickert, L., Reuter, I., Klawonn, F. and Wingender, E. (1998). Transcription regulatory region analysis using signal detection and fuzzy clustering. Bioinformatics, 14(3):244-251.

Polanski, J. and Gieleciak, R. (2003). Comparative molecular surface analysis: a novel tool for drug design and molecular diversity studies. Mol. Divers. 7(1):45-59.

Popescu, M., Gader, P., Keller, J. (2006). Fuzzy spatial pattern processing using linguistic hidden markov models. IEEE Transactions on Fuzzy Systems, 14(1):81-92.

Popescu, M., Keller, J. (2006). Summarization of patient groups using the fuzzy c-means and ontology similarity measures, Proceedings, 15th IEEE International Conference on Fuzzy Systems, Vancouver, Canada.

Popescu, M., Keller, J., Mitchell, J. (2005). Gene ontology automatic annotation using a domain based gene product similarity measure. Proceedings, 14th IEEE International Conference on Fuzzy Systems, Reno, NV, pp. 108-111.

Popescu, M., Keller, J., Mitchell, J., Bezdek, J. (2004). Functional summarization of gene product clusters using gene ontology similarity measures. Proceedings, International Conference on Intelligent Sensors, Sensor Networks and Information Processing, Melbourne, Australia, pp. 553-559.

Popescu, M., Keller, J.M., Mitchell, J.A. (2006). Fuzzy measures on the gene ontology for gene product similarity. IEEE Trans. Computational Biology and Bioinformatics, 3(3):1-11.

Popescu, M., Xu, D., Taylor, E. (2007). GoFuzzKegg: Mapping genes to KEGG pathways using an ontological fuzzy rule system. Proc. of the 2007 IEEE Symposium on Comp. Intell. in Bioinf. and Comp. Biology (CIBCB 2007), Hawaii, USA, pp. 298-303.

Qian, N., Sejnowski, T.J. (1988). Predicting the secondary structure of globular proteins using neural networks. J.Mol. Biol, 202:865-884.

Qiu, H., Keller, J. (1987). Multispectral segmentation using fuzzy techniques. Proceedings, NAFIPS-87, Purdue University, pp. 374-387.

Quackenbush, J. (2002). Microarray data normalization and transformation. Nature Genetics, 32: 496-501.

Renner, S., Ludwig, V., Boden, O., Scheffer, U., Gobel, M., Schneider, G. (2005). New inhibitors of the Tat-TAR RNA interaction found with a "fuzzy" pharmacophore model. Chembiochem.,6(6):1119-25.

Renner, S., Schneider, G.(2004). Fuzzy pharmacophore models from molecular alignments for correlation-vector-based virtual screening. J Med Chem., 47(19):4653-64.

Resnik, P. (1999). Semantic similarity in a taxonomy: an information-base measure and its application to problems of ambiguity in natural language. Journal of Artificial Intelligence Research (JAIR), 11:95-130.

Ressom, H., Reynolds, R., Varghese, R.S. (2003). Increasing the efficiency of fuzzy logic-based gene expression data analysis. Physiol. Genomics, 13:107–117.

Rhee, S.Y., Dickerson J., Xu D. (2006). Bioinformatics and its applications in plant biology. Annu Rev Plant Biol., 57:335-360.

Rost, B., Sander, C. (1993). Prediction of protein secondary structure at better than 70% accuracy. J. Mol. Biol. 232:584–599.

Rost, B., Sander, C. (1994a). Combining evolutionary information and neural networks to predict protein secondary structure. Proteins: Structure, Function and Genetics, 19:55-72.

Rost, B., Sander, C. (1994b). Conservation and prediction of solvent accessibility in protein families. Proteins: Structure, Function and Genetics, 20:216–226.

Rost, B., Sander, C., Schneider, R. (1997). Protein fold recognition by prediction-based threading. J. Mol. Biol. 270:471-480.

Rost. B. (2001). Review: protein secondary structure prediction continues to rise. J. Struct. Biol., 134:204-18.

Russell, B. (1923). Vagueness. Australasian Journal of Philosophy, pp. 88.

Rychlewski, L., Godzik, A. (1997). Secondary structure prediction using segment similarity. Protein Eng, 10:1143–1153.

Sadegh-Zadeh, K. (2000). Fuzzy genomes. Artif. Intell. Med., 18(1):1-28.

Salamov, A.A., Solovyev, V.V. (1995). Prediction of protein secondary structure by combining nearest-neighbor algorithm and multiple sequence alignments. J. Mol. Biol., 247:11-15.

Salamov, A.A., Solovyev, V.V. (1997). Prediction of secondary structure using local alignments. J Mol Biol, 268:31-36.

Sali, A., Blundell, T. L. (1993). Comparative protein modelling by satisfaction of spatial restraints. J. Mol. Biol. 234:779–815.

Samoilov, A., Plyasunov, A. and Arkin, A.P. (2005). Stochastic amplification and signaling in enzymatic futile cycles through noise-induced bistability with oscillations. Proc. Natl. Acad. Sci. USA, 102:2310-5.

Sasik, R., Hwa, T., Iranfar, N. and Loomis, W.F. (2001). Percolation clustering: a novel approach to the clustering of gene expression patterns in Dictyostelium development. Pac. Symp. Biocomput., 335-347.

Schulze-Kremer, S. (1998). Ontologies for molecular biology. In Proceedings of the Third Pacific Symposium on Biocomputing, pp. 693-704.

Sehgal, M.S.B., Gondal, I., Dooley, L., Coppel, R. (2006). AFEGRN- Adaptive fuzzy evolutionary gene regulatory network reconstruction framework, IEEE- World Congress on Computational Intelligence-FUZZ-IEEE, Canada, pp. 8148-8152.

Seising, R. (2005). 1965 - 'Fuzzy Sets' appear – A contribution to the 40th Anniversary, Proceedings, FUZZ-IEEE, pp. 5-10.

Service, R. (2006). The race for the $1000 genome Science, 311:1544-1546.

Shepherd, J.C.W. (1981). Method to determine the reading frame of a protein from the purine/pyrimidine genome sequence and its possible evolutionary justification, Proc Natl Acad Sci, 78:1596-1600.

Shindyalov, I.N., Bourne, P.E. (1998). Protein structure alignment by incremental combinatorial extension (CE) of the optimal path. Protein Engineering, 11(9):739-747.

Sim, J., Kim S., Lee, J. (2005). Prediction of protein solvent accessibility using fuzzy k-nearest neighbor method, Bioinformatics, 21(12):2844–2849.

Sjahputera, O. Keller, J. (2005). Possibilistic C-means, in Scene Matching. Proceedings, Fourth International Conference of the European Society for Fuzzy Logic and Technology (EUSFLAT), Barcelona, Spain, pp. 669-675.

Smith, T. F. and Waterman, M. S. (1981). Comparison of biosequences. Adv. Appl.Math. 2:482–489.

Sokhansanj, B.A., Fitch, J.P., Quong, J.N., Quong, A.A. (2004). Linear fuzzy gene network models obtained from microarray data by exhaustive search. BMC Bioinformatics, 5:108.

Speed T. (2003). Statistical Analysis of Gene Expression of Gene Expression Microarray Data. Chapman & Hall/CRC..

Speer, N., Spieth, C. and Zell. A. (2004). A Memetic Clustering Algorithm for the Functional Partition of Genes Based on the Gene Ontology, Proc. of the 2004 IEEE Symposium on Comp. Intell. in Bioinf. and Comp. Biology (CIBCB 2004), San Diego, California, USA.

Sproule, B.A., Naranjo, C.A., Turksen, I.B. (2002). Fuzzy pharmacology: theory and applications.Trends Pharmacol Sci., 23(9):412-7.

Staden, R., McLachlan, A.D. (1982). Codon preference and its use in identifying protein coding regions in long DNA sequences. Nucleic Acids Research, 10:141-156.

Stanley, R. J., Keller, J., Caldwell, C. W., Gader, P. (2001). Abnormal cell detection using the choquet integral. Proceedings IFSA/NAFIPS, Vancouver, Canada, pp. 1134-1139.

Steinberg, M.S. and McNutt, P.M. (1999). Cadherins and their connections: adhesion junctions have broader functions. Curr. Opin. Cell Biol, 11(5):554-60.

Steuer, R., Kurths, J., Fiehn, O., Weckwerth, W. (2003). Observing and interpreting correlations in metabolomic networks. Bioinformatics, 19(8):1019-26.

Stevens, R., Goble, C.A.,Bechhofer, S. (2000). Ontology-based knowledge representation for bioinformatics. Briefings in Bioinformatics, 1(4):398-416.

Sugeno, M. (1977). Fuzzy Measures and Fuzzy Integrals - A Survey, in Fuzzy Automata and Decision Processes, M. M. Gupta, G. N. Saridis, and B. R. Gaines, eds. New York: North-Holland, pp. 89 - 102.

Sugeno, M. (1985). Industrial Applications of Fuzzy Control. New York: Elsevier Science, Inc..

Sugeno, M., Kang, G.T., (1988). Structure identification of fuzzy model. Fuzzy Sets and Systems, 28(1):15-33.

Sugeno, M., Takagi, T. (1983). Multi-dimensional fuzzy reasoning. Fuzzy Sets Syst., 9:313–325.

Sugeno, M., (1977). Fuzzy measures and fuzzy integrals: A survey, in Fuzzy Automata and Decision Processes, M. Gupta, G. N. Saridis, and B. R. Gaines, eds, Amsterdam, North Holland, pp. 89-102.

Sun, Z., Xia, X., Guo, Q. and Xu, D. (1999). Protein structure prediction in a 210-type lattice model: parameter optimization in the genetic algorithm using orthogonal array. Journal of Protein Chemistry, 18(1):39-46.

Tahani, H., Keller, J. (1990). Automated calculation of non-additive measures for object recognition. Proceedings, SPIE Symposium on Intelligent Robots and Computer Vision IX, Boston, MA, pp. 379-389.

Tahani, H., Keller, J. (1990). Information fusion in computer vision using the fuzzy integral. IEEE Transactions on Systems, Man and Cybernetics, 20(3):733-741.

Takagi. T., Sugeno, M., (1985). Fuzzy identification of systems and its application to modeling and control. IEEE Transactions on System, Man, Cybernetics, 15(1):116-132.

Tanaka, S., Scheraga, H.A. (1975). Model of protein folding: inclusion of short-, medium-, and long-range interactions. Proc Natl Acad Sci USA., 72(10):3802-6.

Taylor, J.A., Johnson, R.S.(2001). Implementation and uses of automated de novo peptide sequencing by tandem mass spectrometry. Anal Chem., 73(11):2594-604.

The Gene Ontology Consortium: http://www.geneontology.org.

Theodoridis, S., Koutroumbas, K. (1998). Pattern recognition, Academic Press, San Diego, CA.

Theodoridis, S., Koutroumbas, K. (2006). Pattern Recognition, 3rd ed., San Diego: Academic Press.

Tho, Q.T., Hui, S.C., Fong A.C.M., Cao T.H. (2006). Automatic fuzzy ontology generation for semantic web. IEEE Trans. Know. Data Eng., 18(6):842-856.

Thompson, J.D., Higgins, D.G., Gibson, T.J. (1994). CLUSTAL W: improving the sensitivity of progressive multiple sequence alignment through sequence weighting, position-specific gap penalties and weight matrix choice. Nucleic Acids Res., 22(22):4673-4680.

Tinoco, I., Jr., Uhlenbeck, O.C., Levine, M.D. (1971). Estimation of secondary structure in ribonucleic acids. Nature, 230(5293):362-7.

Tjhi, W.C., Chen, L. (2006). A partitioning based algorithm to fuzzy co-cluster documents and words. Pattern Recognition Letters, 27(3):151-159.

Toro, J. (2006). Review of "Medical Image Analysis Methods" by Lena Costaridou. Biomed Eng Online, 5(1):6.

Torres, A., Nieto, J.J. (2003). The fuzzy polynucleotide space: basic properties. Bioinformatics, 19(5):587-592.

Torres, A., Nieto, J.J. (2006). Fuzzy logic in medicine and bioinformatics. J Biomed Biotechnol., 2006:91908.

Uddameri, V., Kuchanur, M. (2004). Fuzzy QSARs for predicting logKoc of persistent organic pollutants. Chemosphere., 54(6):771-6.

Verbruggen, H. and Babuska, R. (eds). (1999). Fuzzy Logic Control: Advances in Applications, Singapore: World Scientific.

Vrana, K.E., Freeman, W.M., Aschner, M.(2003). Use of microarray technologies in toxicology research, Neurotoxicology, 24(3):321-32.

Wang, D., Keller, J., Carson, C.A., McAdoo, K., Bailey, C. (1998). Use of fuzzy logic-inspired features to improve bacterial recognition through classifier fusion. IEEE Transactions, Systems, Man, and Cybernetics, 28(4):583-592.

Wang, J., Bo, T.H., Jonassen, I., Myklebost. O. and Hovig, E. (2003). Tumor classification and marker prediction by feature selection and fuzzy c-means clustering using microarray data. BMC Bioinformatics, 4:60.

Wang, L., Larson, E.B., Bowen, J.D., van Belle, G. (2006). Performance-based physical function and future dementia in older people. Archives of Internal Medicine, 166(10):1115-1120.

Wang, T., Keller, J. (2004). Iterative ordering using fuzzy logic and application to ranking college football teams. Proceedings NAFIPS, Banff, Alberta, Canada, pp. 729-733.

Wang, X., Keller, J. (1999). Human-based spatial relationship generalization through neural/fuzzy approaches. Fuzzy Sets and Systems, 101(1):5-20.

Wang, Z. and Klir, G.J. (1993). Fuzzy Measure Theory, Kluwer Academic Publishers, Norwell, MA.

Ward, J.J., McGuffin, L.J., Buxton, B.F., Jones, D.T. (2003). Secondary structure prediction with support vector machines. Bioinformatics, 19:1650-1655.

Washietl, S., Hofacker, I.L, Lukasser, M., Huttenhofer, A., Stadler, P.F. (2005). Mapping of conserved RNA secondary structures predicts thousands of functional noncoding RNAs in the human genome. Nat Biotechnol., 23(11):1383-90.

Waterman, M., and Smith, T.F. (1978). RNA secondary structure: A complete mathematical analysis. Math. Biosc.,42:257-266.

Wilbur, W.J., Lipman, D.J. (1983). Rapid similarity searches of nucleic acid and protein data banks. Proc Natl Acad Sci USA., 80(3):726-30.

Wiswedel, B., Patterson D.E., and Berthold, M.R. (2007). Interactive exploration of fuzzy clusters, in advances in fuzzy clustering and its applications, de Oliveira, J.V. and Pedrycz, W. eds, pp. 123-136, John Wiley and Sons.

Wolstencroft, K., McEntire, R., Stevens, R., Tabanero, L., Brass, A. (2005). Constructing ontology-driven protein family databases, Bioinformatics, 21(8):1685–1692.

Woolf, P.J., Wang, Y., (2000). A fuzzy logic approach to analyzing gene expression data. Physiol. Genomics, 3:9–15.

Xu, D., Bondugula, R., Popescu, M., Keller, J. (2006). Bioinformatics and fuzzy logic, Proceedings, 15th IEEE International Conference on Fuzzy Systems, Vancouver, Canada, pp. 4208-4215.

Xu, D., Xu, Y., Uberbacher, E.C.(2000). Computational tools for protein modeling. Current Protein and Peptide Science, 1:1-21.

Xu, R., Wunsch, D., (2005). Survey of clustering algorithms, IEEE Transactions on Neural Networks, 16(3):645-678.

Xu, Y., Xu, D., Liang, J.(eds.). (2006). Computational Methods for Protein Structure Prediction and Modeling, Volume I & Volume II, Springer-Verlag.

Yager, R. (1980). On a general class of fuzzy connectives. Fuzzy Sets and Systems, 4(3):235-242.

Yager, R. (1988). On ordered weighted averaging aggregation operators in multicriteria decision making. IEEE Trans. On Systems, Man, and Cybernetics, 18(1):183-190.

Yager, R. (2004). Modeling prioritized multicriteria decision making, IEEE Transactions on Systems, Man, and Cybernetics, 34(6):2396-2404.

Yager, R.R. (1983). Quantifiers in the formulation of multiple objective decision functions. Inform. Sci., 31:107–139.

Yager, R.R. (1984). General multiple objective decision making and linguistically quantified statements. Int. J. Man-Machine Studies, 21:389–400.

Yager, R.R. (1988). On ordered weighted averaging aggregation operators in multicriteria decisionmaking. IEEE Transactions on Systems, Man and Cybernetics, 18 (1):183-190.

Yager, R.R. (1993). Families of OWA operators, Fuzzy Sets Syst., 59:125–148.

Yager, R.R. (1996). Quantifier guided aggregation using owa operators. Int. Journal of Intelligent Systems, 11:49-73.

Yager, R.R. (2004). Generalized OWA aggregation operators. Fuzzy Opt. Decision Making, 3: 93–107.

Yang, Y.H., Dudoit, S., Luu, P., Lin, D.M., Peng, V., Ngai, J., Speed, T.P. (2002). Normalization for cDNA microarray data: a robust composite method addressing single and multiple slide systematic variation. Nucleic Acids Res., 30: e15.

Yates, J.R., Eng, J.K., McCormack, A.L. (1996). Mining genomes with MS. Anal. Chem., 67(18):3202-3210.

Yi, T. M., Lander, E.S. (1993). Protein secondary structure prediction using nearest-neighbor methods. J. Mol. Biol., 232:1117-1129.

Zadeh, L. (2002). From computing with numbers to computing with words. Applied Math and Computer Science, 12(3):307-324.

Zadeh, L. (1998). Soft computing, fuzzy logic and recognition technology. Proceedings, IEEE International Conference on Fuzzy Systems, Anchorage, AK, pp. 1678-1679.

Zadeh, L., (1973). Outline of new approach to the analysis of complex systems and decision processes. IEEE Trans. Sys. Man Cyb., 3(1):28-44.

Zadeh, L.A. (1965). Fuzzy sets. Information and Control, 8:338-353.

Zadeh, L.A. (1975a). The concept of a linguistic variable and its application to approximate reasoning, Part 1. Information Sciences, 8:199-249.

Zadeh, L.A. (1975b), The concept of a linguistic variable and its application to approximate reasoning, Part 2. Information Sciences, 8:301-357.

Zadeh, L.A. (1976). The concept of a linguistic variable and its application to approximate reasoning, Part 3, Information Sciences, 9:43-80.

Zemla, A., Venclovas, C., Fidelis, K., Rost. B., (1999). A modified definition of Sov, a segment-based measure for protein secondary structure prediction assessment, Proteins: Structure, Function, and Genetics, 34, 1999, pp. 220-223

Zhang, C.T., Chou, K.C., Maggiora, G.M. (1995). Predicting protein structural classes from amino acid composition: application of fuzzy clustering. Protein Eng. 8: 425–435.

Zhang, X., Mesirov, J.P. and Waltz, D.L. (1992). Hybrid system for Protein Secondary Structure Prediction. J. Mol. Biol., 225:1049-1063.

Zimmermann, H., Zysno, P. (1980). Latent connectives in human decision making. Fuzzy Sets and Systems, 4(1):37-51.

Zuckerkandl, E., Pauling, L. J. (1965). Molecules as documents of evolutionary history. Theor. Biol., 8(2):357-66.

Zurada, J., Marks, R., Robinson, C. (eds) (1994). Computational Intelligence: Imitating Life, New York: IEEE Press.

Index